Rice Crop Science

Rice Crop Science

Sang Yong **Nam** | Kyung-Min **Kim** | Jae-Ryoung **Park** | Jae Hwan **Lee**

Contents

Foreword

Rice is vital to more than half the world's population. It is the most important food grain in the diets of hundreds of millions of Asians, Africans, and Latin Americans living in the tropics and subtropics. Rice will continue to be their primary source of food. Rice can be cultivated throughout our country and is well suited to the climate. It has been passed down from our ancestors and has permeated our lives as farming techniques, traditional proverbs, and traditional learning. Rice is deeply connected to life, religion, and culture beyond being a staple food in our country. It is no exaggeration to say that it is a kind of life itself. Since ancient times, agriculture has been the foundation of the world, and various seasonal customs could be understood only if one knew about rice. In order to cultivate rice well, people were allowed to marry and branch off their families if they understood the farming seasons and the 24 solar terms. Rice has been heaven and a lifeline for generations. When there was a good harvest, the population increased, and when there was a bad harvest, the king was driven out. Even if it meant starving to death due to Boritgogae(period of spring poverty), rice seeds had to be preserved. It was a desperate life, and agricultural techniques were taught to the best of their ability. Its cultivation history is also the oldest among cultivated crops, and it has a history and trajectory that goes back thousands of years. The green revolution that began in the 1960s succeeded with Tongilbyeo in the 1970s, and now, with a self-sufficiency rate of 100%, there is a surplus of rice, and storage warehouses are running out, and due to storage costs, they are considering using it as livestock feed, and many rice varieties for actual feed have been developed. In the meantime, the food self-sufficiency rate has dropped to 23%, and the protein content has dropped below 6% due to the emphasis on taste rather than quantity and the increasing demand for quality. However, as the number of cases of planting other crops has increased due to the alternative crop policy, and the area of rice fields itself has decreased, there are growing voices calling for food security and emergency preparedness. Rice is an excellent food that is easy to digest, tastes good, and provides high-quality energy. It has higher yields and quality than any other crop in the world, and its environmental conservation function, water resource protection, and emotional functions such as landscape are irreplaceable. In addition, with excellent cultivation technology and variety development that has increased the mechanization rate to 98%, there is no great difficulty in actual cultivation, and the related agricultural infrastructure is well-established. These facilities cannot be built overnight, so they are crops that deserve to be protected. We cannot imagine agriculture without rice, nor can we secure happiness or a future without growing rice. Some may argue that it is a simple economic logic, but once the rice production base collapses, it takes much more effort, time, and money to restore it, so we should not judge the core industry that is directly related to life and health by economic logic alone. Rather, we should acknowledge and be grateful that the development of our country's agricultural industry has reached this level due to the devaluation and sacrificial policy of agriculture. Agriculture and the rice industry must be protected and respected as a top priority under any circumstances. Otherwise, we do not know when or how our lives will collapse. We still believe that our abundance and happiness are built on the foundation of the devalued sacrifices of farmers, just like the unconditional sacrifices of our parents. Now, we should be grateful to and respect farmers and pursue a harmonious industrial policy. Agriculture is a happy industry and a life industry, so we must do it even if it means suffering losses, and if we do it, it can become an enjoyable industry. The protection and support policy for agriculture is something that is universally acknowledged around the world. The more developed a country is, the stronger its agriculture is. Nevertheless, it supports and protects. It is true that our country also provides a lot of support, but it still seems insufficient. I would be more than happy if this book were used as a valuable textbook for lectures or civil service exams. There are still many shortcomings. I sincerely hope that you will correct the shortcomings and errors so that it can become a better textbook.

March 3, 2026

From Sang Yong Nam at the foot of Bul-am Mountain

Rice Crop Science

Chapter 1. **Introduction**

Chapter 1.
Introduction

Food crops refer to herbaceous plants, rather than woody plants such as fruit trees or ornamental trees, that are cultivated for consumption with the purpose of obtaining energy. In a narrow sense, food crops excluding horticultural and ornamental crops refer to staple crops, including monocotyledonous cereal crops such as rice, barley, wheat, and maize, as well as dicotyledonous legume crops such as soybeans, peanuts, and peas. Other major food crops include miscellaneous grains like maize and root crops such as potatoes and sweet potatoes. This category also includes industrial crops, forage crops, and green manure crops.

1. Classification and Current Status of Food Crops

As shown in Table 1-2, food crops account for 40% of the world's crop types. Table 1-1 compares the inflorescence and fruit of major crops. The inflorescence (flower) is an important part that determines the breeding and productivity (yield) of the crop. Table 1-5 is a comparative analysis table of monocotyledons and dicotyledons. Table 1-4 shows the scientific name, chromosome number, ploidy, and origin of major crops. Table 1-3 is a table that reclassifies crops by Vavilov origin. If the origin is similar, the climate and soil are similar, which is useful for cultivation as well as pest control. In other words, in cultivation or breeding, the origin of the crop is a good indicator for inferring common growth characteristics and usability.

1) Food Crops

① Rice is a staple food crop, and most of it is paddy rice, with only a small portion being cultivated as upland rice (less than 0.1%).

② Wheat and barley include barley, wheat, oats, and rye.

③ Pulse crops include legumes such as soybean, red bean, mung bean, peanut, kidney bean, and pea.

④ Miscellaneous cereals include maize, millet, barley, foxtail millet, buckwheat, and sorghum.

⑤ Root crops include sweet potato, potato, and other root crops.

2) Industrial Crops

① Starch crops include sweet potato, potato, maize, cassava, etc.

② Oil crops include beans, sesame, perilla, castor-oil plant, rapeseed, sunflower, peanut, flax, cotton, etc.

③ Fiber crops include cotton, hemp, ramie, flax, Indian mallow, tall flat sedge (Cyperus exaltatus), loofah, karinaki, and basket osier.

④ Sugar crops are crops that produce sugar from plants or sap, such as sugar beet, sugar cane, and stevia.

3) Special Crops

① Medicinal crops include mint, ginseng, pyrethrum, hop, etc.

② Stimulant crops include coffee, tobacco, tea, etc.

4) Forage Crops

① Grasses Forage crops include maize, rye, orchard grass, timothy, and ryegrass.

② Legumes include cowpea, red clover, alfalfa, and white clover.

③ Other forage crops include forage pumpkin, turnip, beet, Jerusalem artichoke, and sunflower.

5) Green Manure Crops

① In the Gramineae, there are rye, barley, etc.

② In the Leguminosae, there are purple vetch, clover, hairy vetch, etc.

6) Horticultural Crops

(1) Vegetables

① Stem & leaf vegetables include cabbage, water parsley, lettuce, celery, spinach, asparagus, green onion, onion, garlic, etc.

② Fruit vegetables include pepper, cucumber, watermelon, eggplant, tomato, strawberry, melon, pumpkin, etc.

③ Root vegetables include potato, sweet potato, yam, ginger, radish, carrot, lotus root, burdock, taro, etc.

④ Pod vegetables are the pods of the Leguminosae, including kidney bean, cowpea, pea, and soybean.

(2) Fruit Tree

① Pome fruit is edible when the inflorescence is developed, such as pear, apple, and loquat.

② Drupe fruit is edible when the mesocarp is developed, such as peach, plum, apricot, cherry, jujube, and Japanese apricot.

③ Berry fruit is edible when the epicarp is developed, such as strawberry, grape, and fig.

④ Nut fruit is edible when the cotyledons of the seed are developed, such as chestnut, walnut, and pecan.

⑤ Quasi-pome fruit is edible when the ovary is developed, such as persimmon.

(3) Flowering and Ornamental Plants

① Herbaceous plants include chrysanthemum, dahlia, orchid, marigold, scindapsus, cosmos, etc.

② Woody plants include rose, royal azalea, camellia, rose of sharon (althaea), and weeping fig (*Ficus benjamina*).

Table 1-1. Comparison of the characteristics of the flowering stages and fruits of major food crops

No.	Crop name	Composition and characteristics of inflorescence	Seeds and fruits per inflorescence
①	Barley	1 pistil, 3 stamens	1 seed, several seeds form an ear
②	Maize	Male and female parts separate (monoecious: male tassel, female ear). Female flowers covered with husk	1 seed, several seeds form an ear
③	Potato	1 pistil, 5 stamens, petals and ovary separated	1~4 seeds (1 fruit)
④	Rice	1 pistil, 6 stamens	1 seed, several seeds form an ear
⑤	Soybean	1 pistil, 10 stamens, petals and ovary separate	Multiple seeds (1 pod (column))
⑥	Sweet potato	1 pistil, 5 stamens, petals and ovary separated.	1~4 seeds (1 fruit)

Table 1-2. Types of crops in the world

No.	Crop species	No. of crops	Ratio (%)
①	Dye crops	48	2
②	Fiber crops	97	4
③	Food crops	888	40
④	Forage crops	327	15
⑤	Green manure crops	81	4
⑥	Oil crops	56	3
⑦	Others	128	6
⑧	Seasoning crops	186	8
⑨	Stimulant crops	70	3
⑩	Vegetable crops	342	15
Total		2,223	100

Table 1-3. Classification of crops by region of origin (Vavilov, 1935)

No.	Region	Crops
①	Central Africa (Ethiopia)	Coffee, Melon, Pearl millet, Sorghum, Watermelon
②	Central America (Mexico)	Kidney bean, Maize, Pepper, Pumpkin (Pepo), Sunflower, Sweet potato
③	Central Asian	Apple, Carrot, Fig, Garlic, Millet, Oat, Onion, Pea
④	Chinese (Korean)	Barley (six-row), Buckwheat, Cabbage, Cherry, Ginseng, Milk vetch, Millet, Onion, Peach, Pear, Persimmon, Plum, Red bean, Rice, Soybean, Walnut
⑤	India, Southeast Asia	Banana, Cowpea, Cucumber, Eggplant, Ginger, Gourd, Mung bean, Ramie, Rice, Sesame, Sugarcane
⑥	Mediterranean	Beet, Burdock, Cabbage, Lettuce, Orchardgrass, Pea, Poppy, Rape, Thyme, Timothy, Turnip
⑦	Middle East (Caucasus)	Alfalfa, Apple, Barley (two-row), Fig, Garlic, Grape, Oat, Pear, Quince, Rape, Rye, Spinach, Wheat
⑧	South America (Chile)	Cotton, Kidney Bean, Maize, Peanut, Pepper, Pineapple, Potato, Pumpkin, Tobacco, Tomato

Table 1-4. Scientific names, chromosome numbers, ploidy levels, and centers of origin of major crops

No.	Crop name	Classification	Scientific name	Chromosome No. (2n)	Ploidy	Origin (Vavilov)
①	Alfalfa	Forage crops	*Medicago sativa*	16	2x	Southwest Asia
②	Apple	Fruit trees	*Malus domestica*	34	2x, 3x, 4x	Eastern Europe, Caucasus
③	Barley	Cereal crops	*Hordeum vulgare*	14	2x	Tibet
④	Barnyard millet (edible)	Cereal crops	*Echinochloa frumentacea*	36	4x	India, China
⑤	Broad bean	Legumes	*Vicia faba*	12	2x	North Africa, Southwest Asia
⑥	Buckwheat	Minor cereals	*Fagopyrum esculentum*	16	2x	Northern China
⑦	Cannabis	Special crops	*Cannabis sativa*	20	2x	Central Asia
⑧	Cucumber	Vegetable	*Cucumis sativus*	14	2x	Mediterranean coast
⑨	Eggplant	Vegetable	*Solanum melongena*	24	2x	India
⑩	Flax	Special crops	*Linum usitatissimum*	30	2x	Central Asia
⑪	Grape	Fruit trees	*Vitis vinifera*	38	2x, 3x, 4x	West Asia
⑫	Kidney beans	Legumes	*Phaseolus vulgaris*	22	2x	Central America
⑬	Lettuce	Vegetables	*Lactuca sativa*	18	2x	Mediterranean Coast
⑭	Maize	Cereal crops	*Zea mays*	20	2x	South American Andes
⑮	Millet	Cereal crops	*Panicum miliaceum*	36	4x	Central Asia
⑯	Millet	Cereal crops	*Setaria italica*	18	2x	Eastern Asia
⑰	Oat	Cereal crops	*Avena sativa*	42	6x	Central Asia
⑱	Orange	Fruit trees	*Citrus sinensis*	18	2x	India
⑲	Pea	legumes	*Pisum sativum*	14	2x	Southeast Asia
⑳	Peach	Fruit trees	*Prunus persica*	16	2x	China

㉑	Peanut	legumes	*Arachis hypogaea*	40	4x	Brazil
㉒	Pear	Fruit trees	*Pyrus pyrifolia*(Asian pear) *Pyrus serotina*(China) *Pyrus communis* (European pear)	34	2x	Japan, Southeastern Europe
㉓	Pepper	Vegetable	*Capsicum annuum*	24	2x	South America (Peru)
㉔	Potato	Tuber crop	*Solanum tuberosum*	48	4x	South American Andes
㉕	Pumpkin	Vegetables	*Cucurbita moschata*	20	2x,4x	America
㉖	Radish	Vegetables	*Raphanus sativus*	18	2x	China
㉗	Rape	Vegetable	*Brassica napus*	38	4x	Scandinavia, Caucasus
㉘	Rice	Cereal crops	*Oryza sativa*	24	2x	China, India
㉙	Rye	Cereal crops	*Secale cereale*	14	2x	Caucasus
㉚	Sorghum	Cereal crops	*Sorghum bicolor*	20	2x	Central Africa
㉛	Soybean	Legumes	*Glycine max*	40	2x	Northeastern China
㉜	Strawberry	Vegetables	*Fragaria* x *ananassa*	56	8x	America
㉝	Sugar beet	Industrial crops	*Beta vulgaris*	18	2x	Eastern Europe, Mediterranean
㉞	Sugar cane	Industrial crops	*Saccharum officinale*	80	8x	Southeast Asia
㉟	Sweet potato	Root and tuber crops	*Ipomoea batatas*	90	6x	Central America
㊱	Timothy	Forage crops	*Phleum pratense*	42	6x	Europe, Northern Asia
㊲	Tobacco	Special crops	*Nicotiana tabacum*	48	4x	America
㊳	Tomato	Vegetables	*Solanum lycopersicum*	24	2x	South America Andes
㊴	Watermelon	Vegetables	*Citrullus lanatus*	22	2x	Central Africa
㊵	Wheat	Cereal crops	*Triticum aestivum*	42	6x	Middle East

Table 1-5. Comparison of characteristics of monocotyledonous and dicotyledonous plants

No.	Item	Monocot	Dicot
①	Number of cotyledons	1	2
②	Root shape	root system	taproot system
③	Growth type	herbaceous	herbaceous, woody
④	Shape of leaf vein	parallel vein	reticular vein
⑤	Arrangement of stem vascular bundle	scattered	circular arrangement
⑥	Presence or absence of cambium layer	none (instead, there is an internode growing point)	yes
⑦	Germination holes in pollen	1	3
⑧	Number of petals or stamens	multiples of 3 (rice has 6 stamens, pistil has 1, Y-shaped pistil)  <rice>	multiples of 4 or 5 (potato has 5 petals, 5 stamens, 1 pistil)  <potato>
⑨	Stomata shape	 <corn stomata>	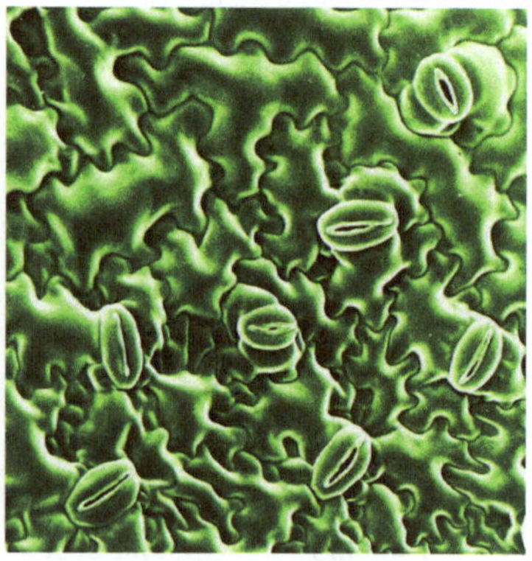 <alfalfa stomata>
⑩	Major plants	rice, barley, lily, wheat, banana, maize	sweet potato, potato, soybean, royal azalea, sunflower

2. Current Status of World Agriculture

1) Classification of Farmland

① Cultivated land refers to land where crops are grown among the total land area.

② Paddy field refers to farmland that is cultivated in a flooded state and mainly grows rice.

③ Upland field refers to farmland other than grassland that is cultivated without flooding (barley, maize, soybean, etc.).

④ Grassland refers to land where annual or perennial grasses are cultivated and used to feed livestock.

2) Status of Land Use Around the World

① The world's land is approximately 13.6 billion ha, and agricultural land (cultivated land and grassland) is approximately 5 billion ha, or approximately 36%.

② The world's cultivated land area is approximately 1.5 billion ha, or approximately 11%.

③ The world's grassland area is approximately 3.5 billion ha, or approximately 25%.

④ Korea has a low grassland ratio of approximately 5%, while the US and UK have high grassland ratios of 20~40%.

⑤ China has a cultivated land area of 130 million ha, or approximately 7% of the world's total.

3) Agricultural Population Around the World

① The agricultural population ratio among the economically active population is 40% worldwide, low in developed countries at around 3%, and high in developing countries at around 50%.

② The agricultural population in Korea and Japan is 4%, in the UK and the US 2%, in Canada and Germany 2.3%, in France 3%, and in China and Vietnam 45%.

4) Production of Major Crops Around the World

① The total world food crop production is about 900 million tons of maize, 700 million tons of wheat, 600 million tons of rice, 200 million tons of barley, 300 million tons of potato, and 100 million tons of sweet potato.

② Looking at the production ratio by region, 90% of the world's rice, 40% of wheat, 40% of potato, and 80% of sweet potato are produced in Asia.

③ Europe produces 90% of rye, 65% of barley, and 45% of potato.

④ About 40% of the world's maize is produced in North America, and about 40% of sorghum is produced in Africa.

5) World Rice Production and Trade

(1) World Rice Production

① There are over 110 rice producing countries, producing 700 million tons on a production area of 150 million ha.

② By continent, Asia accounts for over 90%, America 5.4%, Africa 3.2%, Europe 0.6%, and Australia and the South Pacific 0.1%.

③ According to the Food and Agriculture Organization (FAO) in 2017, the major producing countries are China (150 million tons), India (120 million tons), Indonesia (60 million tons), Bangladesh (35 million tons), Vietnam, Thailand, Myanmar, the Philippines, Brazil, Japan, the United States, and Korea (4 million tons), with Korea ranking 12th.

④ Most of the major production areas are located in areas with high rainfall, and some are produced through irrigation agriculture.

⑤ Temperate *japonica* rice is widely cultivated in Korea, Japan, northern China, California, Egypt, Italy, etc.

⑥ *Indica* rice is mainly produced in southern China, Indonesia, India, Sri Lanka, Thailand, Vietnam, Cambodia, and the Philippines.

⑦ Tropical *japonica* rice is mainly cultivated in southern Indonesia.

⑧ Temperate *japonica* rice accounts for about 22% of China's rice production, and is produced in northern China.

⑨ North Korea's rice production is 2.2 million tons, 58% of South Korea's 3.8 million tons, and the yield is 460 kg/10a, 88% of South Korea's 522 kg/10a.

(2) International Trade of Rice

A: Current status of rice exports and imports

① The world's rice trade volume is only 27 million tons, which is 7% of the production volume, and temperate *japonica* rice is even smaller at around 2 million tons.

② Major exporting countries are Thailand, India, the United States, Vietnam, China, and Pakistan, and major importing countries are Indonesia, Nigeria, the Philippines, Saudi Arabia, Bangladesh, and Brazil.

③ As for the trade volume ratio of international grain trade, wheat and soybean account for 18% of the world's production volume, so they have large exports and imports.

B: Measures to enhance international competitiveness of rice

① Production infrastructure such as rice paddy land consolidation and irrigation and drainage facilities should be improved and production costs should be reduced.

② Differentiation through improvement of distribution system, improvement of labor productivity and land productivity, fostering of full-time rice farmers, and improvement of rice quality is necessary.

③ The multifaceted functions of rice paddy such as environmental conservation should be evaluated to increase production motivation through a direct payment system.

3. Current Status of Agriculture in Korea

1) Proportion of Agriculture

① The share of agricultural production in South Korea's Gross Domestic Product (GDP) has declined from 25% in 1970 to 2.5% in 2020. In comparison, the figure stands at around 10% for China and Vietnam.

② Agricultural income amounts to 53 trillion KRW, while agricultural imports reach approximately 40 trillion KRW. The average annual income per farming household is about 45 million KRW.

③ Within total agricultural income, crop production accounts for 30 trillion KRW. Of this, food crops contribute 10 trillion KRW (with rice making up 8 trillion KRW), vegetables contribute 11 trillion KRW, and fruits 5 trillion KRW.

2) National Land and Agricultural Land Use

① Out of South Korea's total land area of 9.99 million hectares, forests cover 6.3 million hectares (63%), agricultural land covers 1.6 million hectares (16%), and the remaining 19% is used for other purposes. North Korea, with a total area of 12.31 million hectares, has a higher forest area and proportion than South Korea.

② Although the total agricultural land in Korea is decreasing, the proportion of grasslands is increasing, along with rising imports of feed.

③ The proportion of agricultural land to total land has declined from 23% in 1970 to 22% in 1980, 21% in 1990, and further to 16% in 2020.

④ The land use intensity rate (the cultivated area as a percentage of total farmland) was approximately 142 in 1970 due to practices such as double cropping (e.g., barley), but it has dropped to below 100 by 2020.

⑤ The average cultivated area per farming household is about 1.5 hectares, with paddy fields accounting for about 0.9 hectares and dry fields about 0.6 hectares. Although the scale of farming is gradually increasing, it remains relatively small.

3) Agricultural Production in Korea

(1) Food Crops

① In 1970, food crops were cultivated on 2.7 million hectares, producing 6.7 million tons. By 2021, the cultivated area had decreased to 730,000 hectares and production to 3.51 million tons a decline of 83% in area and 48% in output compared to 1970.

② Currently, rice accounts for 88% of total food crop production, followed by naked barley (0.3%), hulled barley (1.4%), malting barley (1.0%), wheat (0.3%), soybean (4%), and maize (2%). This structure shows the overwhelming dominance of rice.

③ Although rice, with a nearly 100% self-sufficiency rate, remains a staple, per capita consumption has decreased from 83 kg in 2000 to 57 kg in 2021 a 30% decline (compared to Japan: 58 kg in 2011 and 53.1 kg in 2018). Meanwhile, due to dietary changes, wheat consumption despite a self-sufficiency rate of around 1% exceeds 37 kg per person annually.

④ As of 2021, the overall grain self-sufficiency rate including feed grains is only 22%. When considering only grains for human consumption, the rate is about 44%.

⑤ Korea imports around 40 trillion KRW worth of agricultural and livestock products, while exports total about 9 trillion KRW, which is roughly 20% of the import value. Processed foods account for approximately 75% of these exports.

⑥ Korea is among the world's top five grain-importing countries, with grain imports (mainly wheat, maize, soybeans, and rice) totaling about 5.3 million tons and costing approximately 2 trillion KRW.

(2) Horticultural Crops

① Most vegetables have seen increases in cultivation area and production, now accounting for around 35% of total agricultural income. Among vegetables, production of strawberries, watermelons, and onions is increasing, while that of traditional vegetables like red peppers is declining. In the case of strawberries, nearly 40% are grown using hydroponics.

② The average per capita vegetable intake is 153 kg about half that of China, but still the second highest in the world. It is also twice the WHO's recommended daily intake (250 g/day).

③ Per capita annual fruit consumption in 2020 was 58 kg, which is 1.6 times higher than Japan's 36 kg. Apple consumption is decreasing, while the diversity of consumed fruits is increasing.

④ In Japan, fruit and vegetable consumption is shifting toward higher quality, greater convenience, and higher added value.

⑤ Among fruit growers in Korea, 6% cultivate persimmons, 3.7% apples, 3.1% peaches, 2.1% grapes, 1.9% citrus, and 1.1% pears.

(3) Livestock

① Livestock accounts for about 42% of total agricultural income (as of 2020), and its importance is increasing. However, aging farmers and environmental issues are worsening the conditions for livestock farming.

② In 2020, the self-sufficiency rates for major livestock products were: beef 36%, pork 77%, chicken 44%, and milk 48%.

③ Due to a shortage of pastures and forage, livestock farming heavily relies on imported feed grains used to produce concentrated mixed feed, including maize and rice straw.

④ Concerns about livestock farming are rising due to the outbreaks of diseases like foot-and-mouth disease and avian influenza, as well as odor and environmental issues.

⑤ Per capita meat consumption in 2020 was 55 kg per year ten times higher than in 1970 including 13 kg of beef and 26 kg of pork.

4) Characteristics of Korean Agriculture

(1) Characteristics of Paddy Field Crops in Korea

① Rice cultivation is the dominant farming system, and agriculture is intensive in nature.

② Cropping systems and pasture-based agriculture are underdeveloped.

③ Soil fertility is low, with organic matter (humus) content around 2%.

④ Korea frequently experiences weather-related disasters such as droughts, cold spells, and typhoons.

⑤ The small scale of farm management leads to weak international competitiveness for agricultural products.

(2) Profitability and Importance of Upland (Field) Crops

① With the westernization of diets, changes in food consumption patterns, and the development of processed and convenience foods, upland crops are gaining a larger share among all food crops.

② Cultivating overwintering crops (e.g., barley) in winter contributes not only to improving food self-sufficiency but also to producing high-quality forage, reducing imports and saving foreign currency.

③ While upland crops typically yield lower income than fruits or vegetables, they play an important role as a stable source of revenue.

④ Many farming machines can be used jointly with rice cultivation, enhancing efficiency and reducing costs.

⑤ Upland crops are essential for balanced diets due to their various functional substances and complementary nutrients (e.g., lysine in rice, methionine in soybean).

⑥ Expanding upland crop cultivation is necessary to improve food self-sufficiency, but many upland fields are located in mountainous or sloped areas, which are unfavorable for farming.

⑦ Upland soils tend to be shallow, with low organic matter and high acidity, often lacking essential nutrients.

⑧ Upland crops are vulnerable to drought in May~June, monsoons and typhoons in July~August, and freezing or dry damage from severe cold in January~February.

⑨ Upland crops have high potential for increased production in emergencies, making them vital for food security.

(3) Challenges Facing Korean Agriculture

① Shift focus from producers to consumers by improving product quality and distribution systems.

② Enhance international competitiveness to respond effectively to globalization and trade.

③ Promote efficient distribution systems and improve storability to maintain product quality.

④ Increase productivity by improving both yield per unit area and product quality.

⑤ Diversify and upscale product quality, including taste, functionality, and processability.

⑥ Diversify crop species and cultivation methods to meet regional and market-specific demands.

⑦ Pursue sustainable agriculture through the establishment of eco-friendly practices, reducing inputs while ensuring safety.

⑧ Strengthen agricultural exports to prevent overproduction and stabilize farmers' income.

⑨ Overcome production limitations and promote market stability through overseas agricultural development.

5) Agricultural Population and Farm Management Scale

① The number of farming households and the agricultural population has been continuously decreasing. As of 2021, there are about 980,000 farming households and approximately 2.15 million people engaged in farming.

② The proportion of the agricultural population has significantly declined, from 45% in 1970 to 4.5% in 2020.

③ As of 2020, full-time farming households account for about 60% and part-time farming households about 40%.

④ In terms of farm size, 52% of households manage less than 0.5 hectares, 22% manage 0.5~1 hectare, 14% manage 1~2 hectares, and 9% manage over 2 hectares, with the average farm size being only 1.1 hectares indicating a generally small-scale farming structure.

⑤ The average age of farm owners is 66, with 50% aged 65 or older. Most have over 30 years of farming experience, and 50% have completed high school or higher education.

6) Agricultural Land Fertility

① Korean paddy soils are categorized as follows: normal paddy fields (32%), sandy paddy fields (31%), immature soils (23%), wetland paddies (9%), and saline soils (3%). Low-productivity paddies account for about two-thirds of the total.

② For upland fields, the distribution is: normal fields (42%), sandy soils (23%), immature soils (17%), clay soils (14%), and volcanic ash soils (2%), with low-productivity fields making up 58% of the total.

③ In terms of soil chemistry, the soil pH is 5.8 for paddy fields, 6.2 for upland fields, and about 6.4 for greenhouse cultivation.

④ Organic matter content (humus) is approximately 2% in both paddy and upland soils, which is below the target level of 3%.

⑤ Available phosphorus (P) levels are: 130 mg/kg in paddy fields, 680 mg/kg in upland fields, 670 mg/kg in orchards, and 1,000 mg/kg in greenhouse cultivation all exceeding the appropriate range.

⑥ The cation exchange capacity (CEC) for major fertilizer nutrients like potassium (K), calcium (Ca), and magnesium (Mg) is around 10 cmol/kg, which is relatively low. However, orchard and greenhouse soils tend to have higher CEC values.

⑦ The available silicon (Si) content in paddy fields is about 75 mg/kg, which is below the optimal range of 130~180 mg/kg, indicating the need for silicate fertilizers. Silicate fertilizer is the most commonly used fertilizer in rice paddies.

7) Analysis and Enhancement of Farm Household Income

① In Korea, the average annual farm household income is about 45 million KRW. Of this, agricultural income accounts for around 12 million KRW (27%), non-agricultural income 17 million KRW (38%), transfer income 14 million KRW (31%), and non-regular income approximately 3 million KRW (5%).

② Of the total agricultural income, crop cultivation accounts for 73%, while non-crop sources contribute about 27%.

③ Within gross agricultural income, the share of rice has declined to 17%, while income from vegetables, fruits, and specialty crops has increased to 40%, and livestock to 43%.

④ Farm household income increases proportionally with farmland size. However, to reach the income level of urban workers, farms must operate on at least 7 hectares.

⑤ Based on agricultural product sales: 19% of farms earn less than 1.2 million KRW, 44% earn between 1.2 and 10 million KRW, 17% earn between 10 and 30 million KRW, and 13% earn over 30 million KRW.

8) Management Analysis of Rice Culture

(1) Labor and Production Costs

① In 2002, the labor input for rice culture in Korea was 28.6 hours per 10a. In the U.S., due to large-scale culture (around 200 ha in the 1970s), labor input had already decreased to 2~3 hours per 10a.

② As of 1997, the labor investment time by cultivation method is approximately 30 hours for mid-rice transplanting, 27 hours for young seedling transplanting, 22 hours for direct seeding in dry fields, and 23 hours for average direct seeding.

③ As of 2022, the labor time by work stage is 5 hours for transplanting, 5 hours for water management, 5 hours for seedling cultivation, 2 hours for combine harvesting, 2 hours for pest and disease contro.

④ Labor time is slightly higher in Gyeongbuk, Jeonbuk, and Gyeongnam than the national average.

⑤ To improve international competitiveness of rice, labor-saving technologies and mechanization must be promoted, and labor input per operation stage must be reduced.

⑥ The mechanization rate of rice culture in Korea is very high, reaching 99%, on the other hand, upland crops is about 67% in 2023.

(2) Rice Production Costs

A: Rice production costs in Korea

The rice production cost is 970,000 won per 10a in 2020. The rice revenue was 1.21 million won, and the net profit after deducting the production cost was 440,000 won, with a net profit rate of 36%. The income after deducting the management cost of 480,000 won from the revenue was 730,000 won, with an income rate of 60%. The breakdown of rice production costs by item was land service cost of 270,000 won, capital service cost of 10,000 won (indirect production cost of 280,000 won), labor cost of 190,000 won, farming cost of 30,000 won, pesticide cost of 30,000 won, and fertilizer cost of 50,000 won.

① Land service cost

To lower rice production costs, the largest cost component land service cost must be reduced. Korean farmland prices are higher than in the U.S. or Europe. Expanding farm size and improving land use efficiency are essential.

② Machinery cost

Although the mechanization rate for rice farming is 98%, improving efficiency requires scaling up operations, maintaining agricultural infrastructure to extend machine life, and implementing comprehensive mechanization strategies and cropping system reforms.

③ Labor cost

While labor costs have declined due to labor-saving technologies, machinery-related costs such as fuel have increased. It is necessary to develop technologies that reduce both labor and machinery costs simultaneously.

④ Pesticide cost

Although pesticide usage has decreased due to eco-friendly trends, high prices have kept pesticide costs relatively stable. Solutions include developing disease- and pest-resistant varieties, improving nutritional resistance, and implementing integrated nutrient and pest management and precision agriculture.

⑤ Fertilizer cost

With growing awareness of eco-friendly farming, fertilizer usage is declining. To ensure safe and high-quality rice production, use of organic fertilizers and soil conditioners should be increased.

B: Rice production costs in Japan

The cost of rice production in Japan (2017) is 1.17 million won per 10a, of which 31% is labor cost, 22% is irrigation cost, 10% is rent, 8% is fertilizer, 7% is pesticide, and 22% is other costs. The labor cost is about 7% higher than in Korea, and the land service cost (rent) is about 25% lower. At this time, the income is 2.4 million won, so the income rate is about 50%.

C: Rice production costs in major countries

According to the 2017 Rural Development Administration data including land service cost, the rice production cost in Korea is 920,000 won/10a, and when this is considered 100, the US is 45%, China is 48%, Taiwan is 66%, and Japan is 121%. The increase in global rice production has increased in proportion to the development and use of chemical fertilizers and pesticides.

D: Strategies to reduce rice production costs

① Achieve cost-effectiveness through expansion of farm size (economies of scale).

② Improve cost structure through efficient use of agricultural machinery (leasing or scaling).

③ Enhance farm management and production capabilities.

④ Introduce labor-saving cultivation technologies.

⑤ Select high-yield varieties and improve cultivation practices to increase productivity per unit area.

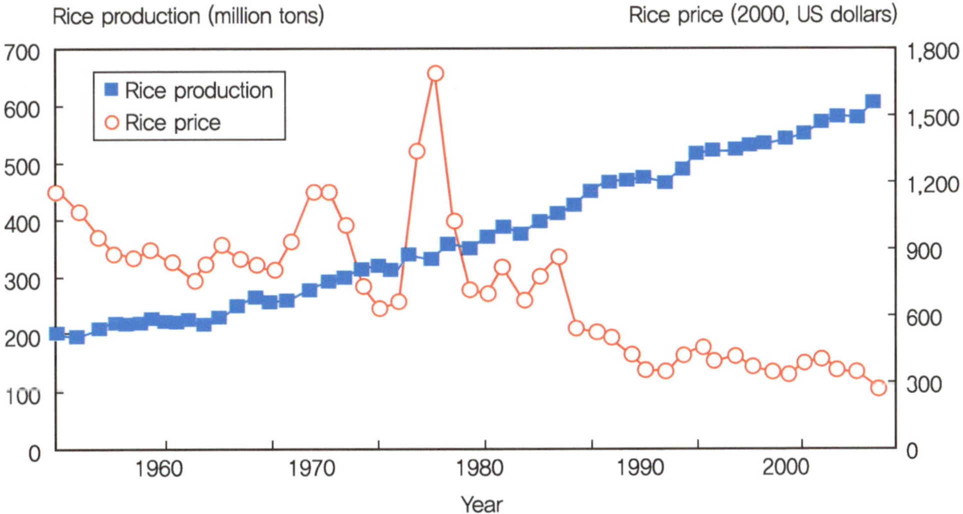

Figure 1-1. World rice production and price trends

<Slender-shaped: *Indica* rice> <Short-rounded: *Japonica* rice>

Figure 1-2. Differences between *Indica*(left) and *Japonica* (right) rice seeds
* In Indica rice, the lemma and palea are tightly adhered to the caryopsis.

4. Classification of Rice and Current Status in Korea

1) Name and Classification of Rice

(1) Name of Rice

A: The scientific name of cultivated rice is *Oryza sativa* L.

① The Chinese character for rice, "Do, 稻", which resembles the shape of pounding rice in a mortar, is used in Korea, China, and Japan.

② Rice before milling is called rough rice, and after milling is referred to as polished rice. Regardless of milling status, the general term milled grain is used to refer broadly to rice.

③ In Korea, rice is typically stored in its husked (rough) form, while in Japan, it is stored as brown rice. Yield is now commonly converted and calculated in terms of polished rice rather than rough rice.

④ Rice cultivated in paddies is called water rice, while rice cultivated in dry fields is called upland rice. In Korea, rice is almost exclusively grown in paddies.

B: Botanical classification of rice

① Rice belongs to the family Gramineae (Poaceae), genus *Oryza*, which contains 24 species. The main cultivated species are the Asian rice (*Oryza sativa*) and the African rice (*Oryza glaberrima*).

② Most rice cultivated worldwide is the Asian species, while the African species is grown in limited areas of West Africa, primarily as upland rice.

③ Rice is a diploid plant with 2n=24 chromosomes, possessing a pair of 12 chromosomes (n=12). Its genome composition is AA.

④ Chromosome analysis of rice has revealed that its genome consists of approximately 40,000 genes.

C: Classification of cultivated rice

① The Asian rice (*Oryza sativa*) developed from a perennial wild rice, *Oryza rufipogon*, into an annual cultivated species.

② The African rice (*Oryza glaberrima*) evolved from a perennial wild rice known as *Oryza barthii* or *Oryza breviligulata* into an annual cultivated form.

③ *Oryza sativa*, the Asian rice species, is further classified into three ecotypes, which are temperate *japonica*, tropical *japonica*, and *indica*.

D: Comparison between cultivated and wild rice

Cultivated rice has a lower outcrossing rate, larger seed size, weaker seed shattering, and little to no dormancy compared to wild rice. It also exhibits stronger resistance to lodging and weaker responses to photoperiod and temperature changes, but has lower tolerance to cold. On the other hand, wild rice, often regarded as weedy rice, possesses strong resistance to diseases, pests, and environmental stresses. As such, it is considered a valuable genetic resource and a key subject in ecological research.

Table 1-6. Classification of crops based on fertilization type

Self-pollinated crops	Barley (*Hordeum vulgare*), Eggplant (*Solanum melongena*), Grape (*Vitis vinifera*), Pea (*Pisum sativum*), Peach (*Prunus persica*), Rice (*Oryza sativa*), Sesame (*Sesamum indicum*), Soybean (*Glycine max*), Tobacco (*Nicotiana tabacum*), Tomato (*Solanum lycopersicum*), Wheat (*Triticum aestivum*)
Cross-pollinated crops	Asparagus (*Asparagus officinalis*), Buckwheat (*Fagopyrum esculentum*), Chinese Cabbage (*Brassica rapa* subsp. pekinensis), Garlic (*Allium sativum*), Hop (*Humulus lupulus*), Job's Tears (*Coix lacryma-jobi*), Maize (Corn, *Zea mays*), Onion (*Allium cepa*), Pepper (*Capsicum annuum*), Radish (*Raphanus sativus*), Rye (*Secale cereale*), Spinach (*Spinacia oleracea*), Strawberry (*Fragaria × ananassa*)

Table 1-7. Comparison of seed and physiological-ecological differences between wild and cultivated rice

No.	Trait category	Item	Cultivated rice	Wild rice
①	Seed traits	Number of seeds	Many	Few
		Seed longevity	Short	Long
		Seed size	Large	Small
		Seed shattering	Low	High
		Dormancy	Weak	Strong
②	Reproductive traits	Time to anther dehiscence after flowering	Immediate	Around 30 minutes
		Mode of fertilization	Self-pollinating (outcrossing rate ~1%)	Cross-pollinating (outcrossing rate 30–99%)
		Stigma size	Small	Large
		Pollen dispersal distance	20 m	40 m
		Pollen number per anther	700–2,500 grains	4,000–9,000 grains
		Pollen viability duration	3 minutes	6 minutes
③	Physiological traits	Photoperiod sensitivity	Sensitive to insensitive	Sensitive
		Temperature sensitivity	Sensitive to insensitive	Sensitive
		Cold tolerance	Weak	Strong
		Lodging resistance	Strong	Weak
④	Ecological traits	Awn characteristics	Absent or short	Long and sturdy
		Reproduction mode	Sexual (seed propagation)	Sexual and vegetative propagation
		Life span	Annual	Annual or perennial
		Panicle shape	Long and compact	Short and spreading

Table 1-8. Chromosomes and genome composition of rice (Lee, H.S., 1998)

Group	Chromosome number	Genome composition	Species name	Geographic distribution
I	12	AA	*Oryza sativa* L.	Asia
			Oryza glaberrima Steud.	West Africa
			Oryza breviligulata	West Africa
			Oryza perennis (*Oryza rufipogon*)	South Asia, Southern China
		EE	*Oryza australiensis*	Australia
II	24	BBCC	*Oryza minuta* Presl.	Southeast Asia
III	24	CCDD	*Oryza latifolia* Desv.	Central America, South America

2) Origin and Dissemination of Rice

The origin of rice is estimated using archaeological methods such as historical records and excavation of ancient sites, as well as molecular biological techniques including enzyme analysis and DNA analysis. Most scholars agree that rice species began to diverge approximately 10,000 years ago.

(1) Centers of Origin of Rice

A: Assam region of India theory

① According to Vavilov's theory of centers of origin, the Assam region in India is considered the most probable origin of rice.

② Supporting this theory is the discovery of carbonized rice dated to around 5,000 B.C. in this region.

③ Additionally, pottery containing rice grains dated to around 6,000 B.C. has been excavated from nearby areas.

B: Thailand and Vietnam theory

① In this region, artifacts presumed to be related to rice cultivation have been excavated from Neolithic cultural sites dating back approximately 10,000 years.

C: Yunnan, China and Myanmar theory

① Wild rice still grows naturally in abundance in this region, and the high level of genetic variation supports the idea that this may be a center of origin. Many scholars support this theory, although archaeological evidence remains insufficient.

D: Southern China theory

① Wild rice continues to grow naturally in South China, and many rice cultivation-related artifacts have been found in this region.

② Carbonized rice dated to around 9,000 B.C. has been excavated from this area.

(2) History of Rice Cultivation

① Based on the combined results of radiocarbon dating of artifacts and carbonized rice found around the world, rice cultivation is estimated to have begun approximately 10,000 years ago.

② In Korea, the history of rice cultivation is inferred from ancient texts and archaeological findings as follows:

ⓐ The Records of the SamgukjiWijidongyijeon (삼국지 위지동이전), published over 2,600 years ago, contains references to rice.

ⓑ The Chronicles of Baekjebongi (백제서기) and Samguk Sagi (삼국사기) includes records of rice farming, suggesting it began around 2,000 years ago.

ⓒ The discovery of carbonized rice grains in Goyang City, Gyeonggi Province in 1991 is estimated to date back approximately 5,000 years.

(3) Routes of Rice Dissemination

There are four hypothesized routes by which rice was introduced to the Korean Peninsula.

① From the Yellow River basin in China, rice spread along the Chinese coast via land, through the Yalu River region, and into Korea.

② From the Shandong Peninsula in China, it traveled through the Liaodong Peninsula, north of the Yellow Sea, into Korea.

③ From the Shandong Peninsula, it crossed the Yellow Sea by sea and reached the lower Han River basin.

④ From the Yellow River basin, it crossed the Yellow Sea by sea and spread into the southern regions of the Korean Peninsula.

3) The Value and Importance of Rice

(1) Climatic and Ecological Perspectives

① The staple crop of a region is typically the one best adapted to its climatic and ecological conditions, and should be producible in the most economically efficient manner.

② In arid regions or areas unsuitable for grain cultivation, livestock farming becomes predominant, giving rise to meat-based dietary cultures.

③ In the warm and dry regions of the Americas, wheat and maize have become the dominant food and energy crops due to their high adaptability.

④ In semi-arid temperate zones, drought-tolerant upland crops such as wheat and barley are easy to cultivate, leading to a bread-based dietary culture. Wheat's adaptation to cold and dry climates contributed to the development of western civilizations.

⑤ In monsoon-dominated temperate regions, where seasonal rainfall is abundant, rice became the staple food due to its adaptation to flooded environments. Rice has evolved as a high-temperature, high-humidity crop suited to paddy cultivation systems.

⑥ In Korea, rice demonstrates optimal adaptation to local agro-climatic conditions, achieving high yields and storability, making it the most important staple crop cultivated and consumed throughout the year.

⑦ Unlike rice, crops such as maize, potato, and sweet potato introduced to Korea more recently have remained secondary due to lower compatibility with national preferences, cultural identity, and lower storability.

⑧ Upland crops such as barley, wheat, maize, and soybean are more susceptible to monsoons and typhoons, making them less reliable than rice in terms of production and utilization under Korea's climate.

(2) Cultural and Societal Perspectives

① The three major staple crops globally are rice, wheat, and maize. While wheat occupies the largest cultivation area, maize ranks highest in total production. Rice production is comparable to both, with each yielding approximately 600 million tons annually.

② In Korea, rice cultivation is deeply embedded in cultural identity and collective consciousness. This is partly because paddy farming is perceived as more manageable than upland farming.

③ Unlike wheat farming, rice cultivation often requires cooperative labor, which has contributed to the development of strong community cohesion and social capital in rural areas.

④ Rice is not only agro-ecologically well-suited to Korea but has also played a key role as an economically viable staple crop. Its cultivation has shaped a national character associated with wisdom, perseverance, and collective effort.

(3) Energy and Nutritional Perspectives

① From an energy standpoint, the caloric value per 100g of edible grain is: rice 360 kcal, wheat 358 kcal, and maize 362 kcal. The edible portion (edibility ratio) is rice 75%, wheat 62%, and maize 84%.

② The population-supporting capacity per hectare per year is estimated at rice 22 persons, wheat 14 persons, and maize 19 persons highlighting the suitability of rice for densely populated regions like Asia.

③ Rice protein content is about 7%, lower than wheat (12%), but its amino acid profile is superior in biological value and digestibility for human nutrition.

④ Although rice is relatively low in certain amino acids such as lysine, it is nutritionally well-balanced and approaches the profile of a complete food, with proven benefits in preventing various lifestyle-related diseases.

Table 1-9. Composition of major food crops (Yamaguchi 1978)

No.	Crop	Protein(%)	Fat(%)	Carbohydrates(%)	Ash(%)
①	Rice	7.8	2.7	87.0	1.5
②	Barley	11.6	2.2	83.4	2.8
③	Wheat	12.1	2.3	83.7	1.9
④	Maize	9.5	5.3	83.7	1.5
⑤	Soybean	39.0	19.9	35.4	5.7
⑥	Potato	9.2	0.5	86.3	3.9
⑦	Sesame	21.2	54.7	18.4	5.7

*The protein content of rice in Korea is currently being reduced to below 6% in consideration of eating quality (palatability).

(4) Food Security Perspective

A: Staple crop production in Korea

① Among Korea's staple food crops, only rice, potato, and sweet potato are produced at levels that meet domestic self-sufficiency, while the cultivated area and production volume of other staple crops remain very limited.

② Although potato and sweet potato are relatively easy to cultivate, they are difficult to store, making them unsuitable as long-term staple foods.

③ Wheat, maize, and soybean are economically uncompetitive due to the influx of cheaper imported products.

④ In Korea, rice is virtually the only crop that allows for stable domestic production and supply.

⑤ Since Korea has very limited pastureland for livestock grazing, the country heavily depends on imported feed, making it difficult to claim true domestic origin for livestock products.

B: Food security and rice (Paddy/Rice grain)

① Food security refers to the ability to consistently procure sufficient quantities of food of acceptable quality for the population, at the right time and place, and to sustain this state over the long term.

② The core elements of food security include availability, accessibility, stability, and safety.

③ Livestock products that rely heavily on imported feed are largely disconnected from the framework of national food security.

④ Rice is a crop that satisfies multiple food security criteria, including alignment with national policy objectives, energy efficiency, and high self-sufficiency potential.

4) Multifunctional Roles of Rice Cultivation

In addition to its primary role of producing food, agriculture provides a range of non-market public goods such as balanced regional development, agricultural employment, environmental conservation, food security, and the preservation of traditional culture. In recognition of these multifunctional benefits, both Korea and other countries implement direct payment schemes funded by public finances to support the agricultural sector.

Table 1-10. Environmental conservation functions of rice cultivation

No.	Category	Description	Estimated economic value
①	Air purification	During photosynthesis, rice plants absorb carbon dioxide (CO_2) and release oxygen (O_2), thereby contributing to atmospheric purification. This dual function CO_2 absorption and O_2 release is recognized as a key ecosystem service of crop cultivation.	KRW 1.74 trillion
②	Air temperature regulation	Water in flooded paddy fields evaporates or transpires through rice plants, effectively cooling the surrounding atmosphere during hot periods. In colder seasons, deep water storage warmed by solar radiation can buffer temperature drops, protecting rice plants from cold stress. In Korea, evapotranspiration from paddy fields totals approximately 4.5 billion tons annually.	KRW 1.36 trillion
③	Biodiversity conservation	Paddy fields and associated irrigation channels serve as important habitats for a wide range of aquatic and terrestrial organisms, supporting biodiversity conservation in agricultural landscapes.	—
④	Flood regulation	Paddy fields function as temporary water reservoirs during heavy rainfall, contributing significantly to flood mitigation. The total water storage capacity of all paddy fields in Korea is approximately 3.6 billion tons.	KRW 11.4 trillion
⑤	Groundwater recharge	Water held in paddy fields continuously infiltrates into the subsoil. Of the infiltrated water, approximately 55% flows into rivers, while 45% equivalent to 15.8 billion tons is recharged into groundwater aquifers. This contributes to both groundwater replenishment and baseflow maintenance in river systems.	KRW 1.4 trillion
⑥	Soil erosion control	Paddy fields intercept and retain soil that would otherwise be lost through surface runoff on sloped upland fields. If converted to dryland conditions, paddy fields can lose up to 22 tons of soil per hectare annually. Maintaining flooded conditions helps prevent soil erosion and conserves topsoil.	KRW 200 billion
⑦	Water purification	When paddy fields are irrigated with wastewater from industrial and domestic sources, rice plants absorb nitrogen (N), while phosphorus (P) is removed through plant uptake and soil fixation. Nitrogen applied to paddy soil remains largely in the ammonium (NH_4^+) form and is adsorbed to soil particles, reducing runoff risk. Nitrogen is purified via plant uptake, soil denitrification, and nitrogen fixation by algae. In ordinary paddy fields, nitrogen removal efficiency reaches 25~40% during early growth stages and exceeds 90% at the heading stage.	KRW 1.9 trillion

In Korea, the direct payment system includes programs for public-interest functions in agriculture, forestry, and fisheries. Since 2001, Korea has implemented a direct payment program specifically for paddy farming (Rice Paddy Direct Payment Program).

Note: “Mu-non” refers to permanently flooded paddy fields. “Mo” denotes rice seedlings. “Iyang” refers to the traditional transplanting of rice seedlings, a term that became standardized in Korean due to its long-standing colloquial usage.

5) Rice Cultivation as the Foundation of Agricultural Production

(1) The Importance of Rice in Korean Agriculture

① The area under rice cultivation accounts for approximately 10% of the national land area and about 50% of the total arable land.

② Households cultivating rice represent about 77% of all farming households in Korea.

③ Rice production constitutes approximately 88% of the total domestic output of food crops.

④ For rice-growing households, rice contributes to 30% of total household income and 76% of their agricultural income.

(2) Reasons for the Continued Prevalence of Rice Cultivation

① Crops such as wheat, maize, and soybean are economically unviable due to their low price and quality competitiveness, prompting farmers to shift to high-value crops such as specialty crops, vegetables, fruits, or livestock production.

② Alternative crops to rice are constrained by small-scale operations and limited market demand, and even slight increases in production can lead to market oversupply.

③ Rice farming is highly mechanized, with a farm mechanization rate of approximately 98%, making rice cultivation and management more efficient compared to other crops.

(3) Current Status of Rice Production and Supply in Korea

A: Production and consumption overview

① In 2022, Korea cultivated rice on 730,000 hectares (ha) of paddy land, producing approximately 3.51 million metric tons, achieving nearly 100% self-sufficiency in rice.

② Per capita annual rice consumption declined significantly from 136 kg in 1970 to 58 kg in 2020.

③ The total annual rice demand is approximately 3.7 million tons, while the domestic production in 2021 reached 3.8 million tons.

④ The overall food self-sufficiency rate is merely 23%, and as of September 2021, the national rice reserve was 150,000 tons far below the optimal reserve level of approximately 800,000 tons.

⑤ The FAO (Food and Agriculture Organization of the United Nations) recommends maintaining a rice reserve equivalent to 17~18% of annual consumption.

B: Tongilbyeo and the history of rice supply and demand

① Prior to the 1970s, Korea faced chronic rice shortages.

② In 1971, the high-yielding Tongilbyeo variety was developed and disseminated, enabling rice self-sufficiency by 1977.

③ By 1978, Tongilbyeo was cultivated on 78% of all paddy fields nationwide.

③ In 1980, a major cold weather event severely reduced yields from large-scale Tongilbyeo cultivation, leading to an emergency import of 1.4 million tons of rice.

④ With rising consumer preference for high-quality rice, the adoption of Tongil-type rice began to decline from 1989 onward.

⑤ By 1992, Tongilbyeo cultivation had been entirely phased out, and consumer demand shifted toward high-quality and functional rice varieties.

⑥ In 2007, rice varieties for feed use were developed and trialed.

C: Area and yield of rice cultivation

① In 2020, the total area under rice cultivation was 726,000 ha, and the yield per 10 ares (at a milling ratio of 92.9%) was 483 kg.

② When applying a milling ratio of 90.4%, the adjusted yield per 10 ares becomes 470kg.

③ Based on the 92.9% milling rate, total production in 2020 was 3.51 million tons, and applying the 90.4% milling rate yields an adjusted production of 3.41 million tons.

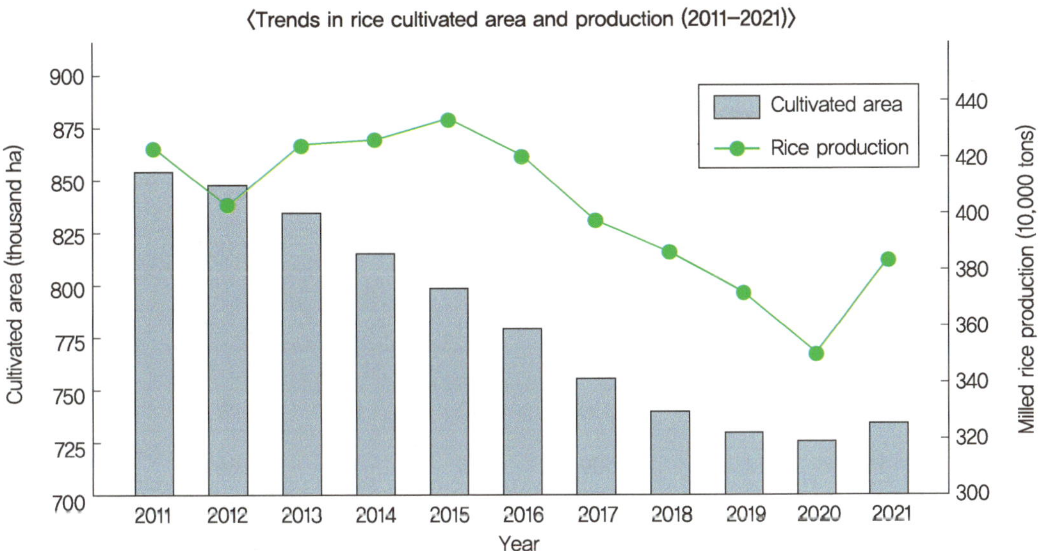

Figure 1-3. Trends in annual rice cultivated area and milled rice production

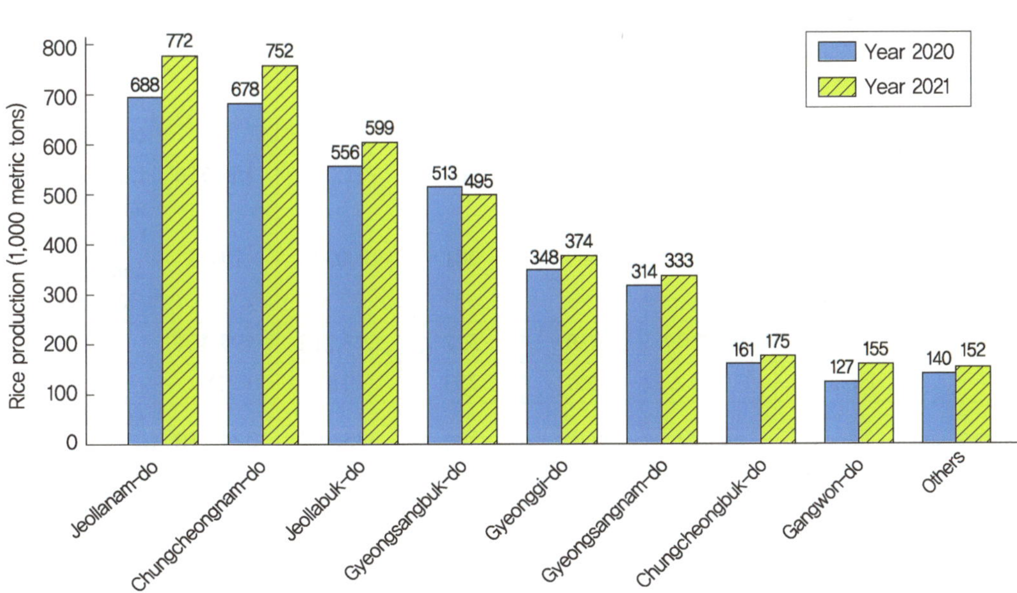

Figure 1-4. Rice production by province (Unit: 1,000 metric tons) (Note: "Others" includes special metropolitan cities, Sejong special self-governing city, and Jeju special self-governing province)

Table 1-11. Rice cultivation area and production in South Korea by province (2021)

Province	Cultivation type	Cultivated area (ha)	Yield per 10a (kg)		Total production (ton)		Remarks
			9/10* milling	12/10* milling	9/10 milling	12/10 milling	
Seoul	Paddy	187	501	488	937	912	No upland rice cultivation
Busan	Paddy	2,169	525	511	11,387	11,081	"
Daegu	Paddy	2,847	510	496	14,514	14,123	"
Incheon	Paddy	11,747	496	483	58,296	56,727	"
Gwangju	Paddy	4,910	494	481	24,265	23,612	-
	Upland	6	300	292	19	18	
Daejeon	Paddy	1,131	509	495	5,756	5,601	No upland rice cultivation
Ulsan	Paddy	3,793	491	478	18,624	18,123	"
Sejong	Paddy	3,340	550	535	18,351	17,857	"
Gyeonggi	Paddy	74,717	501	488	374,373	364,298	"
Gangwon	Paddy	28,903	535	521	154,513	150,355	"
Chungbuk	Paddy	33,403	525	511	175,338	170,620	"
Chungnam	Paddy	135,398	555	540	752,046	731,808	-
	Upland	1	315	307	2	2	-
Jeonbuk	Paddy	114,509	523	509	599,330	583,202	No upland rice cultivation
Jeonnam	Paddy	155,101	597	484	771,333	750,576	Largest area and yield from both paddy and upland rice cultivation
Jeonnam	Upland	334	309	301	1,031	1,003	
Gyeongbuk	Paddy	95,830	536	522	513,212	499,401	-
	Upland	7	260	253	18	18	-
Gyeongnam	Paddy	64,079	520	506	333,073	324,110	No upland rice cultivation
Jeju	Paddy	6	334	525	18	18	More upland rice than paddy rice cultivated
	Upland	59	285	277	168	163	
National total (17 regions)	Paddy	732,070	523	508	3,825,367	3,722,424	Cultivated area and production show a declining trend year over year
	Upland	407	304	296	1,239	1,206	

*In the statistical data, the 9/10 milling (9-bundo) figures correspond to a milling yield of 92.9%, whereas the 12/10 milling (12-bundo) figures correspond to a milling yield of 90.4%.

D: Future rice demand forecast

① Assuming an annual per capita rice consumption of 60kg and a maximum population of 50 million, a minimum of 3 million metric tons of rice will be required annually.

② Considering inter-Korean unification, if we assume 70kg of rice consumption per capita and a total population of 70 million, the annual demand for rice would increase to 5 million metric tons.

③ Based on an average paddy yield of 500kg per 10 ares (equivalent to 5 tons per hectare), to produce 3 million metric tons of rice, approximately 600,000 hectares (ha) of paddy fields would be needed.

④ From a food security perspective, to prepare for potential contingencies, the indiscriminate conversion of paddy land to non-agricultural use must be restrained, and the current cultivated area of 700,000 hectares (ha) should be maintained.

Rice Crop Science

Chapter 2. **Rice Morphology and Structure**

Chapter 2.
Rice Morphology and Structure

Rice (*Oryza sativa*) is a cereal crop belonging to the *Poaceae* family, cultivated for its edible caryopsis. A caryopsis is a type of dry fruit that functions as a seed and is characteristic of grasses. Among cultivated crops, few are as terminologically ambiguous, broadly defined, and variably classified as rice. In Chapter 2, we aim to systematically organize and understand the terminology related to the approximately 150-day growth cycle of rice, beginning with the seed and progressing through the root, leaf, and stem stages, culminating again in seed formation. The portion of rice consumed as food is the seed, specifically the grain or kernel. The complexity and diversity of rice morphology reflect its cultural and societal importance.

First and foremost, the seed must be well-developed and viable. The root system plays a critical role in anchoring the plant and ensuring efficient uptake of water and nutrients. The culm (stem) must support numerous leaves, arranged in a spatially efficient manner to maximize photosynthetic efficiency. The leaves must be physiologically adapted for optimal performance of photosynthesis, respiration, and transpiration.

Finally, the floral organs and the panicle structure of rice possess characteristics that are well-suited to the climatic and soil conditions of Korea. It is essential to study and understand the anatomical structures, terminologies, morphological forms, and functional roles of these components in order to gain a comprehensive understanding of rice as a crop species.

1. Seeds of Rice

1) Characteristics of Rice Grain

① The seed of rice is botanically classified as a caryopsis. this spikelet is a small unit of the panicle.

② In grasses (*Poaceae family*), the seed of rice is specifically referred to as a caryopsis.

③ Brown rice is the form in which the caryopsis is covered by a hull, which is the outer protective layer of the fruit.

④ The hull consists of two components: the lemma (outer glume) and the palea (inner glume). The tip of the lemma often bears an awn, a bristle-like appendage.

⑤ *Japonica*-type rice is typically short-grained, round, and glutinous (sticky), whereas *Indica*-type rice is long-grained with a lower degree of stickiness.

2) Structure of the Rice Seed

① The grain is anatomically composed of several layers from the outside inward: the pericarp, seed coat, aleurone layer, and outer endosperm.

② The pericarp corresponds to the hull (removed in milling), and the mesocarp (middle layer) is equivalent to the bran layer in brown rice.

③ Brown rice consists of the embryo (2n), endosperm (3n), and seed coat (2n). The embryo is made up of the plumule (embryonic shoot), hypocotyl, and radicle (embryonic root).

④ The plumule includes the apical meristem, and is protected by the coleoptile, which encloses the developing primary leaves. The primordia of the first to third true leaves are already differentiated.

⑤ The radicle includes the seminal root, which is protected by the coleorhiza, and both are already differentiated at seed maturity.

3) Internal Structure of the Rice Grain

① The internal structure of the grain consists of the outer endosperm, aleurone layer, and starch cells.

② In non-glutinous rice (ordinary rice), the starch cells contain approximately 20% amylose, resulting in a translucent appearance.

③ In contrast, glutinous rice (sticky rice) exhibits a milky-white and semi-translucent appearance due to the presence of micropores within the starch cell structure, which cause diffuse reflection of light. Its amylose content is generally around 10% or less.

④ The parenchyma tissue consists of thin primary cell walls and is capable of photosynthesis and cell division. The collenchyma tissue is characterized by thickened primary cell walls and functions as a supportive tissue in the aboveground parts of dicotyledonous plants.

⑤ The sclerenchyma tissue serves as a protective tissue in monocotyledonous plants, composed of fibers and sclereids with lignified secondary cell walls, providing both mechanical strength and elasticity.

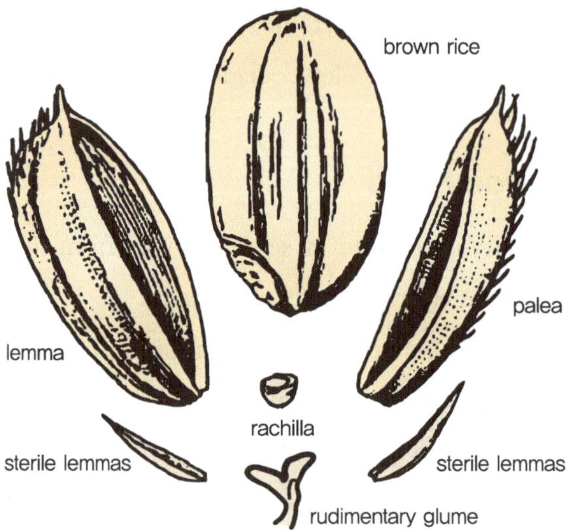

Figure 2-1. Structure of the rice grain (Hoshikawa, 1975)

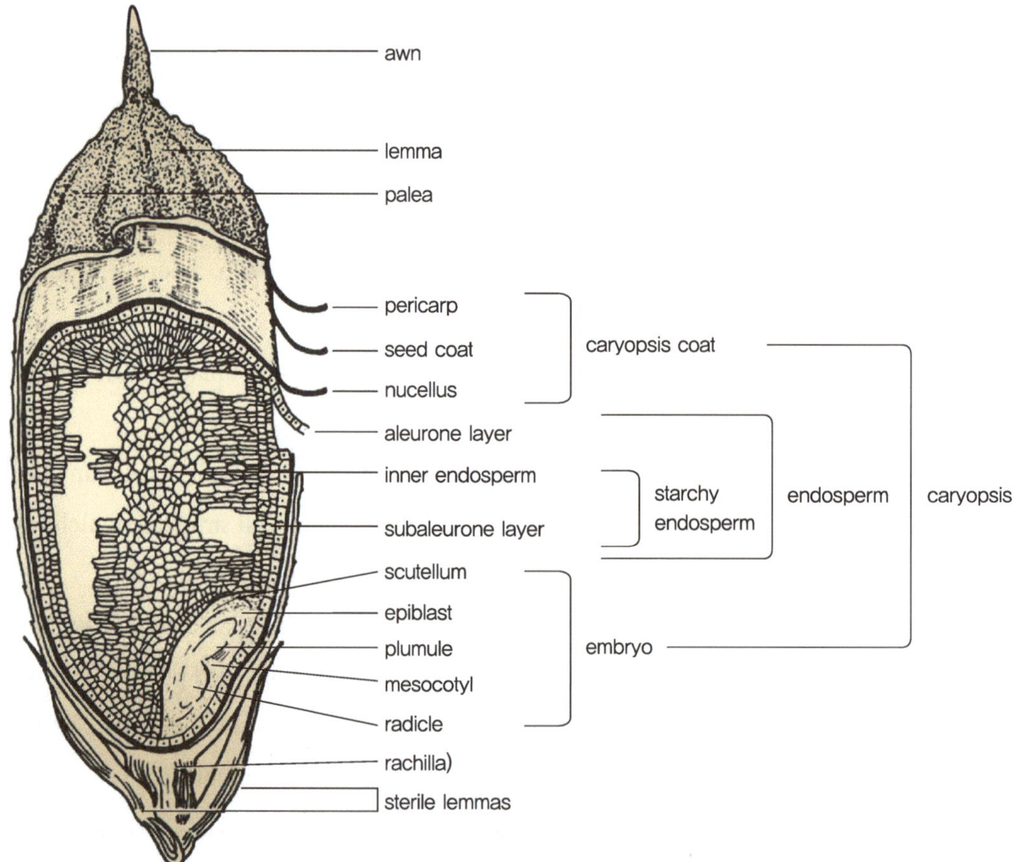

Figure 2-2. Cross-section of a rice grain (Juliano and Aldama, 1937)

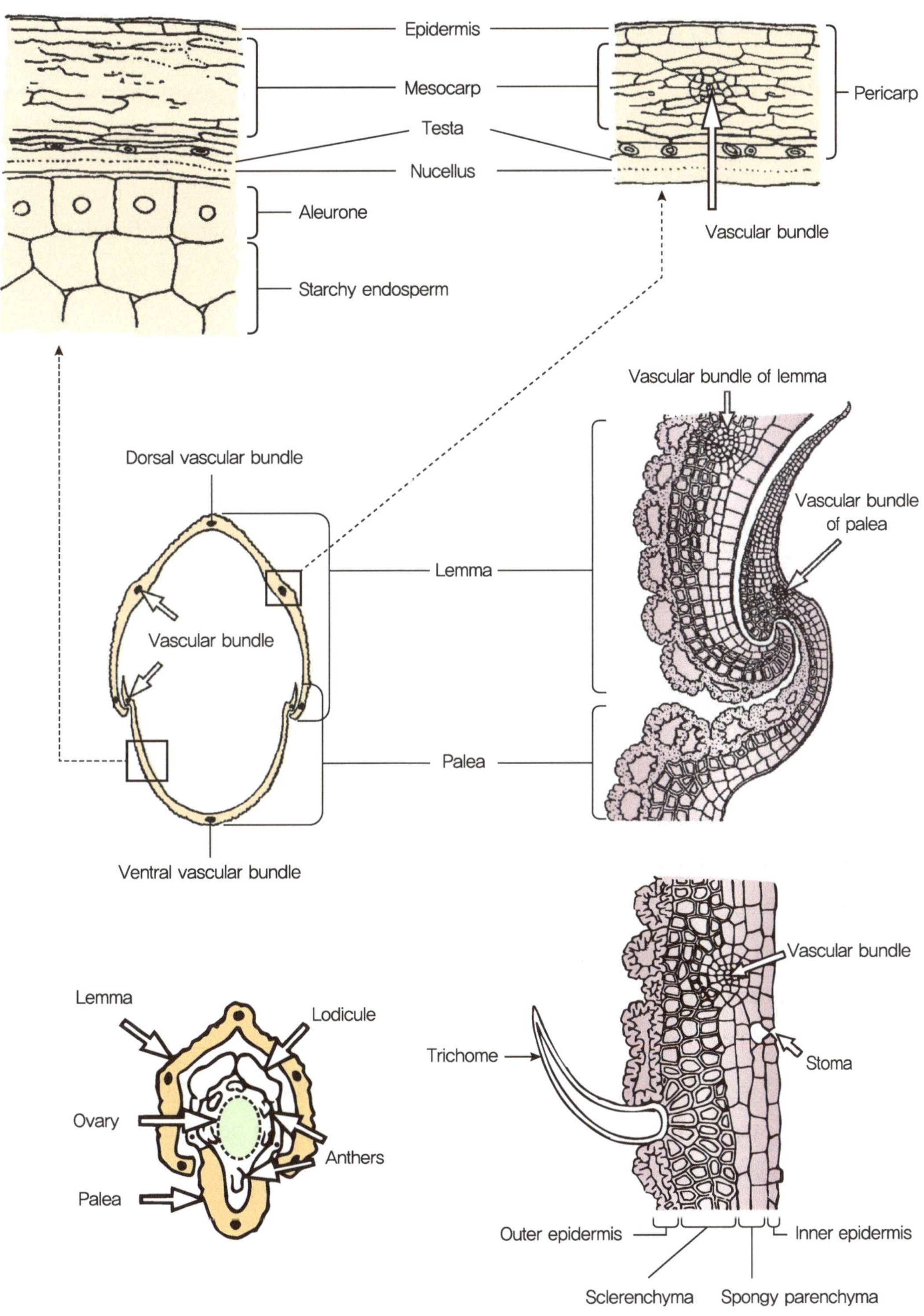

Figure 2-3. Epidermal tissue of the rice grain

2. Rice Root System

1) Seminal Root

① The seminal root is the primary root that elongates first during germination, emerging through the coleorhiza from the seed, and can grow up to 15cm in length.

② It absorbs water and nutrients immediately after germination and remains functional until the 7-leaf stage, even after crown roots have developed.

2) Mesocotyl Root

① This is an adventitious root that does not typically emerge under normal sowing conditions but elongates when the seed is sown too deeply.

② When rice seeds are sown too deep in dry seedbeds or in direct dry seeding systems, mesocotyl roots frequently develop.

3) Crown Root

① Crown roots are adventitious roots that emerge from the stem nodes and constitute the main part of the rice root system. They are also called nodal roots because they originate from each node. Around each node, 10 to 20 roots are formed circumferentially.

② The primary roots that emerge directly from the node are the thickest, and they give rise to secondary and tertiary roots which become progressively thinner, with the thinnest being the branch roots.

③ The number of crown roots per node increases toward the upper nodes, reaching a peak at the 11th node (when up to the flag leaf, 15 leaves have developed), and then decreases at nodes above this level.

④ The most vigorous root elongation and lateral root development occur during the panicle initiation stage. After flag leaf emergence, the root system becomes net-like and is concentrated near the soil surface.

4) Root Hair

① Root hairs are slender, thread-like extensions from the epidermis near the root tip, which play a crucial role in absorbing water and nutrients from the soil.

② They are well-developed in the epidermal cells of the maturation zone and are supported by a cortex composed of parenchyma cells.

③ In mature rice plants, root hairs degenerate, and the epidermal cells become lignified (corkified). This lignified layer, along with the suberized inner sclerenchyma cells, provides protection to the root surface.

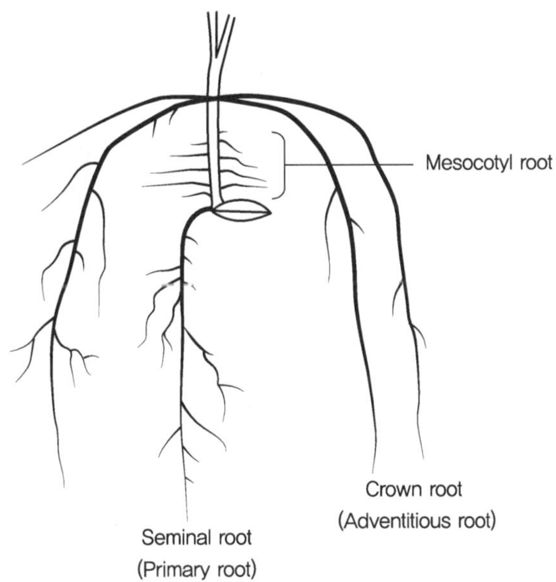

Figure 2-4. Three types of rice roots (Hoshikawa, 1975)

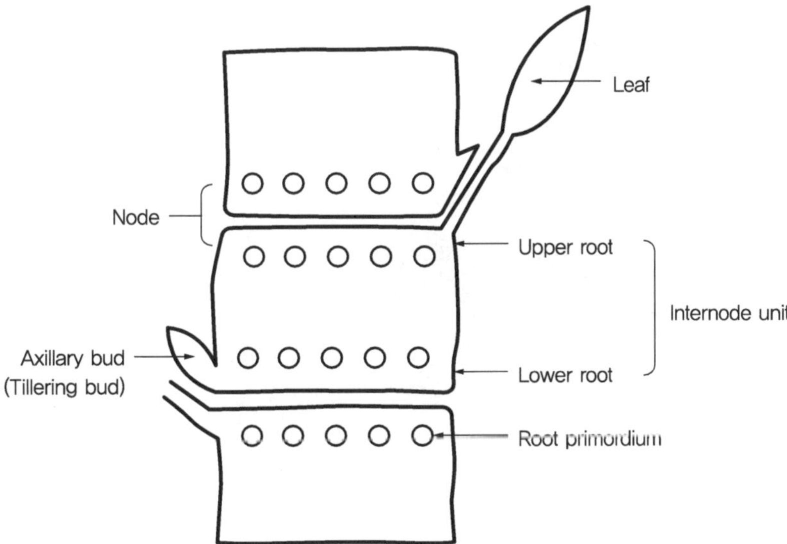

Figure 2-5. Structural units of leaf, root, and tillering (Kawataet al., 1963)

5) Lysigenous Aerenchyma

① As rice plants grow, the cortex of the root undergoes programmed cell death, leading to the formation of air spaces and the development of lysigenous aerenchyma.

② These air spaces connect with the aerenchyma in the leaves and stems, forming a continuous pathway for oxygen transport from the aerial parts to the roots.

③ The supplied oxygen is utilized in root cellular respiration under the anaerobic conditions of paddy soils and protects the roots from toxic gases such as hydrogen sulfide.

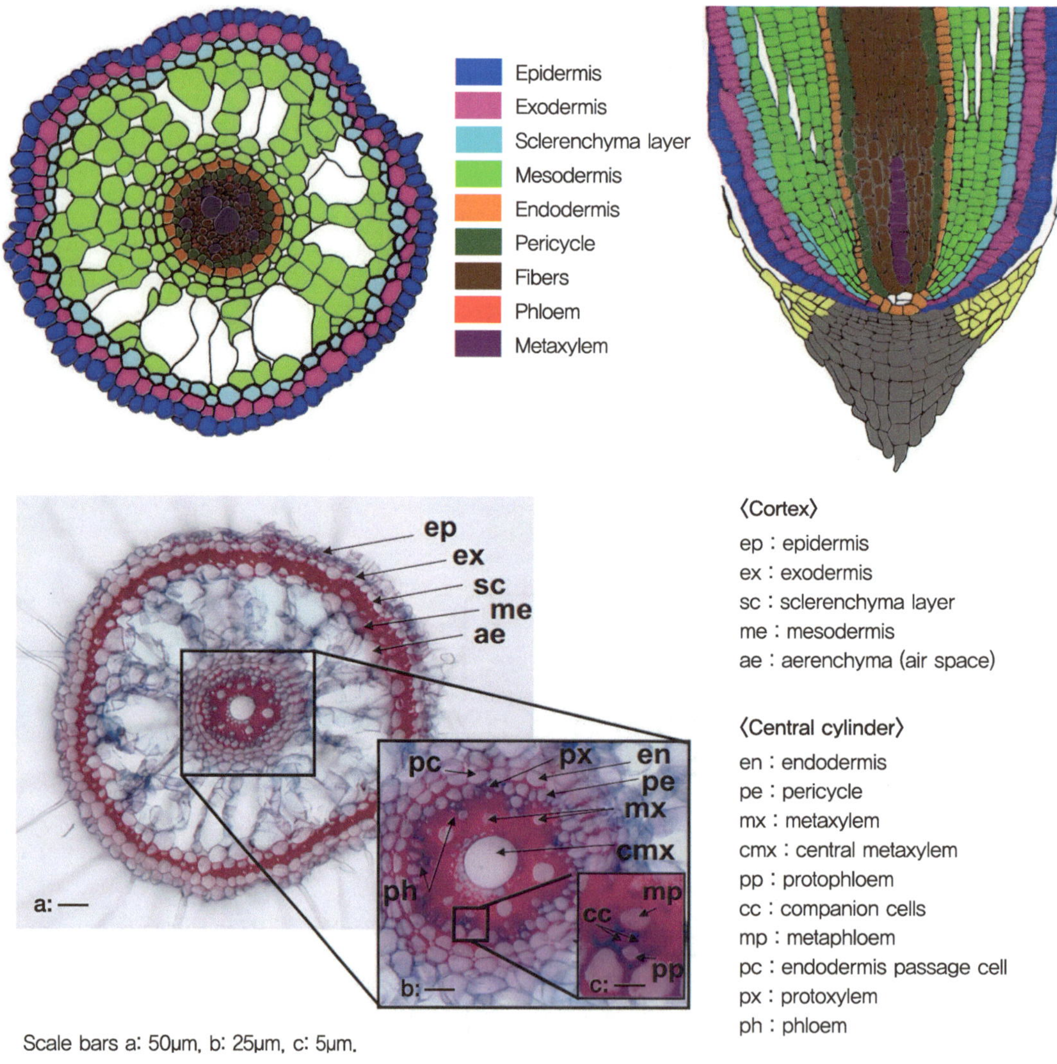

Figure 2-6. Radial anatomy of radicle (Rebouillat et al., 2009)

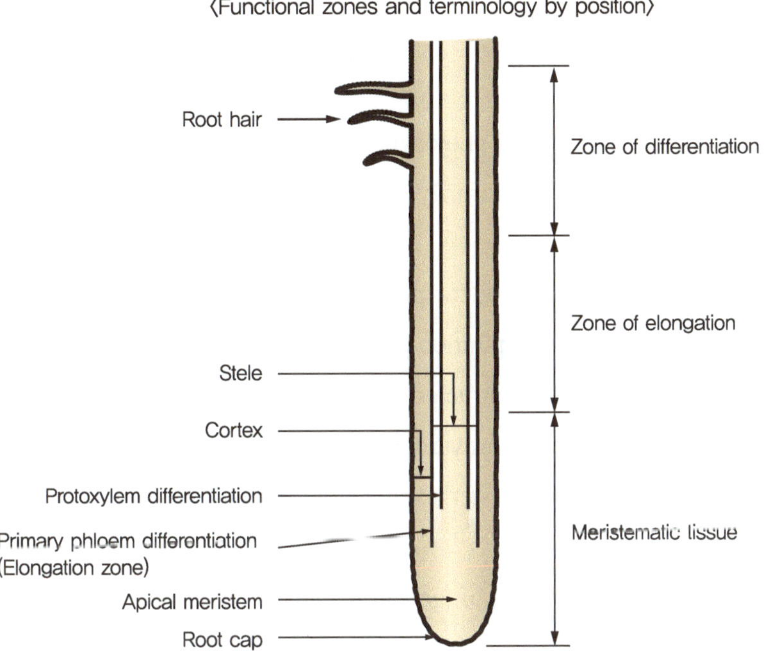

Figure 2-7. Structure and function of a young root (Kramer, 1969)

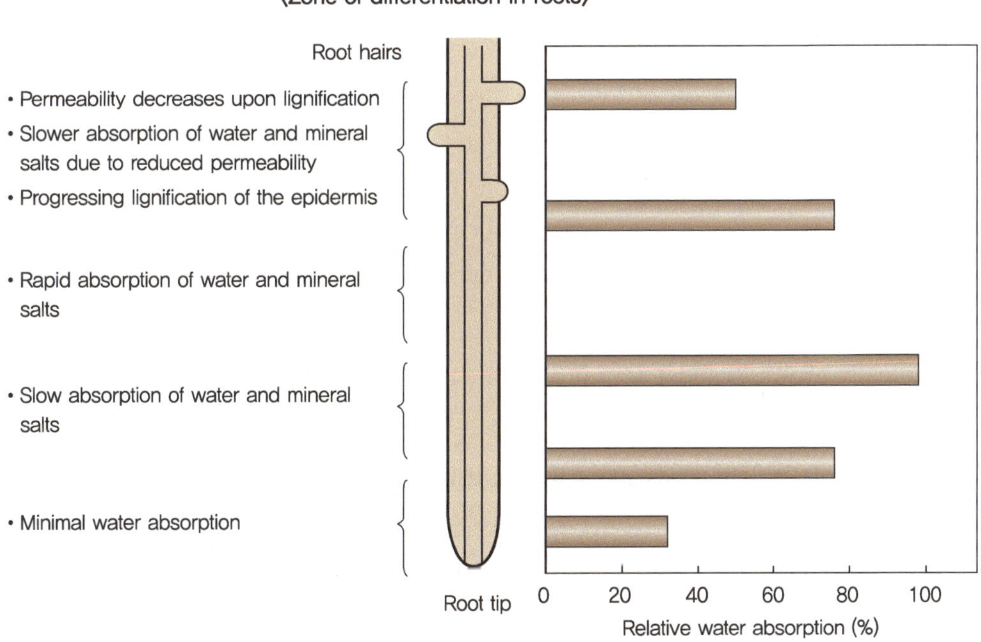

Figure 2-8. Water absorption by different root zones in rice (Kawata and Lai, 1968)

3. Leaves

1) Structure of Leaf

① The leaf sheath performs minimal photosynthesis but functions to protect the internodal leaf and the young panicle, temporarily store assimilates, wrap around the stem, and prevent lodging.

② The leaf blade is the primary site for photosynthesis and transpiration. Motor cells (bulliform cells) regulate the transpiration process.

③ The leaf ligule is a tongue-shaped structure that is developmentally part of the leaf sheath. It is present in rice but absent in barnyardgrass (Echinochloa). It prevents rainwater from entering between the sheath and stem and helps regulate air humidity during drying conditions.

④ The auricle is developmentally part of the leaf blade and helps keep the sheath attached to the stem.

⑤ Although some rice cultivars may lack ligules and auricles, most rice plants possess ligules, whereas barnyardgrass does not. Therefore, the presence or absence of the ligule and auricle serves as a key characteristic for distinguishing between rice and barnyardgrass.

2) Types of Leaves

(1) Coleoptile

① The coleoptile is the first leaf to emerge during germination, serving to protect the true leaves of the young shoot.

② It typically grows to about 1cm in length and is not considered a true leaf, so it is not included in the leaf count.

③ When the coleoptile reaches about 1cm in length in a germinating rice seed, the first true leaf begins to emerge.

④ The seminal root also begins to emerge with the coleoptile. The coleoptile can emerge even under low oxygen conditions.

(2) First Leaf (1st True Leaf)

① It has a cylindrical shape and appears as a pointed needle-like leaf due to the underdevelopment of the leaf blade.

② The first leaf lacks a leaf blade and grows only about 2cm of leaf sheath, so it is considered an incomplete leaf.

③ Even though it is incomplete, it is counted in the total leaf number.

3) Second Leaf (2nd True Leaf)

① The second leaf has a short, narrow, spoon-shaped leaf blade.

② It emerges before the first leaf is fully developed. The sheath of the second leaf grows to about 3~5cm, and its spoon-shaped blade grows about 2cm, which is shorter than the sheath.

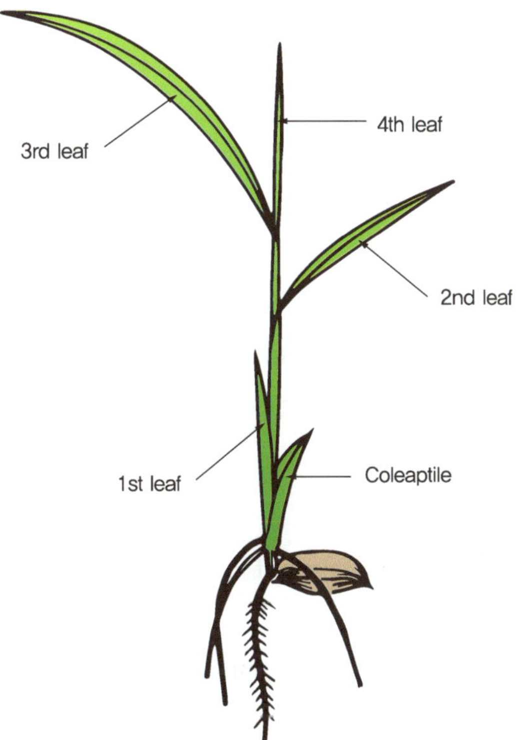

Figure 2-9. Number of leaves (3.5 leaves) and nomenclature in rice

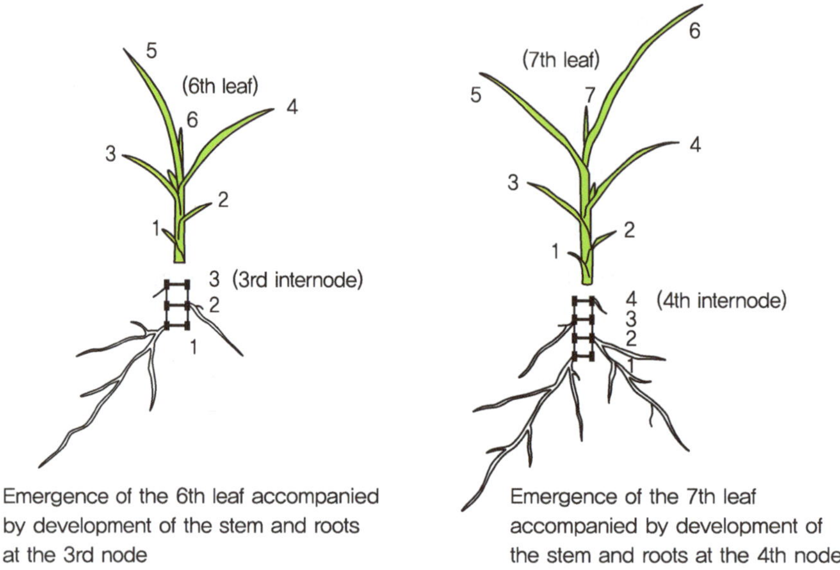

Figure 2-10. Diagram of the synchronous growth pattern of tillers, leaves, and roots (Fujii, 1974)

(4) Third True Leaf (Third Leaf)

① All leaves that emerge after the third true leaf exhibit a fully developed leaf structure. The leaf sheath of the third leaf grows to approximately 4~6cm, and the leaf blade extends to about 5~7cm.

② The third and subsequent leaves emerge from the top of the sheath of the preceding fully developed leaf. From the third leaf onward, the leaf blade becomes longer than the leaf sheath.

(5) Flag Leaf (Terminal Leaf)

① The terminal leaf is the final uppermost leaf to emerge. It is shorter in length but wider in width compared to the leaf just below it.

② The flag leaf encloses the young panicle (immature inflorescence) within its blade prior to heading and remains active until the end, performing photosynthesis.

③ It emerges approximately 18 days before heading, and around 6 days prior to heading, the sheath of the flag leaf swells due to the developing panicle inside a stage referred to as the booting stage.

(6) Prophyll

At the time of tillering, the first leaf to appear lacks a blade. This leaf corresponds to the coleoptile of the main shoot and functions to protect the tiller bud.

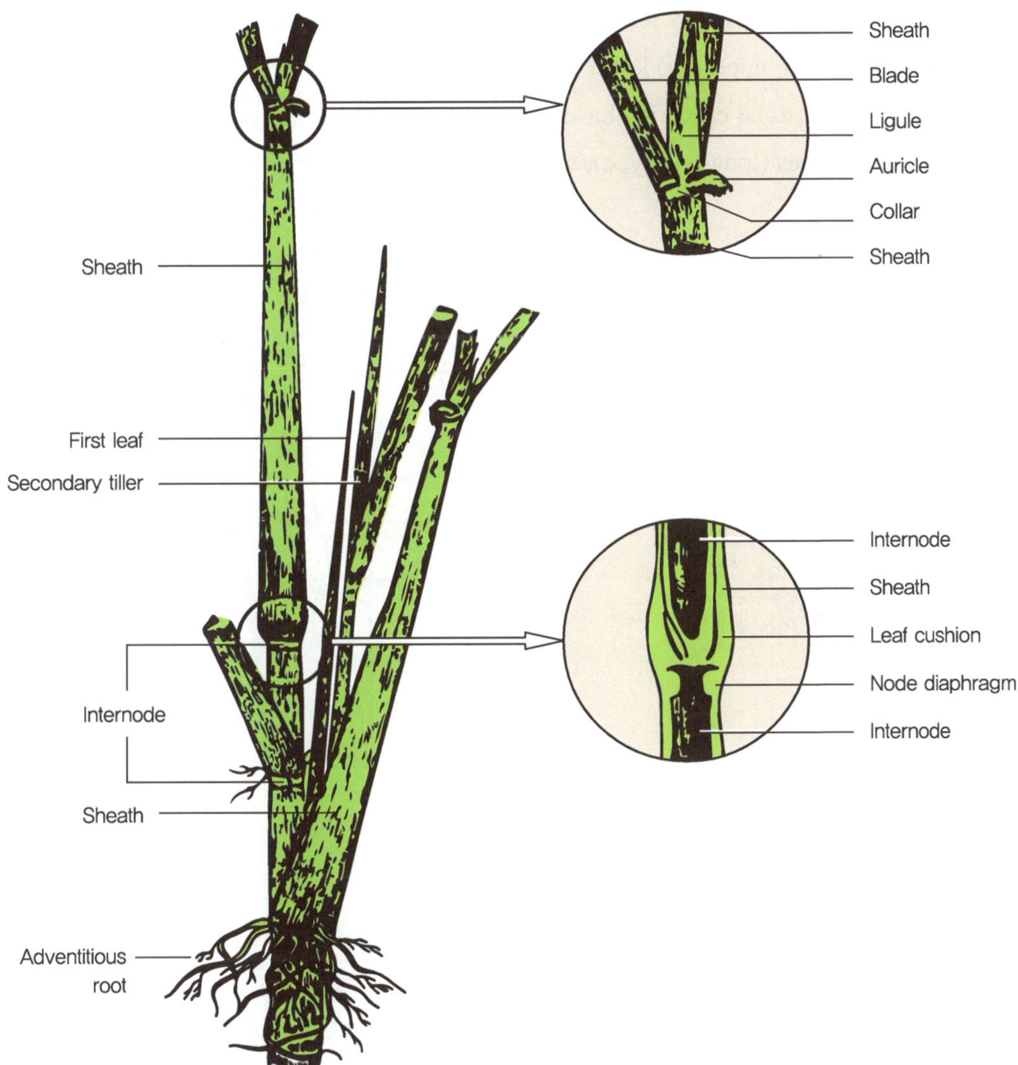

Figure 2-11. Structure of tillers and nodes (Chang and Bardens, 1965), https://labs.plb.ucdavis.edu/rost/rice/Leaves/external.html. https://www.dreamstime.com /photos-images/oryza-longistaminata.html

3) Leaf Characteristics

① The number of leaves varies by cultivar. For rice cultivated in Korea, early-maturing varieties typically produce around 14 leaves per main culm, late-maturing varieties produce about 18, with an average of approximately 16 leaves.

② Leaf blade length varies by leaf position. The third leaf below the flag leaf is usually the longest, with both upper and lower leaves becoming progressively shorter.

③ Cultivars can also be classified based on leaf blade morphology: short leaves (short-leaf type), long leaves (long-leaf type), wide leaves (broad-leaf type), narrow leaves (slender-leaf type).

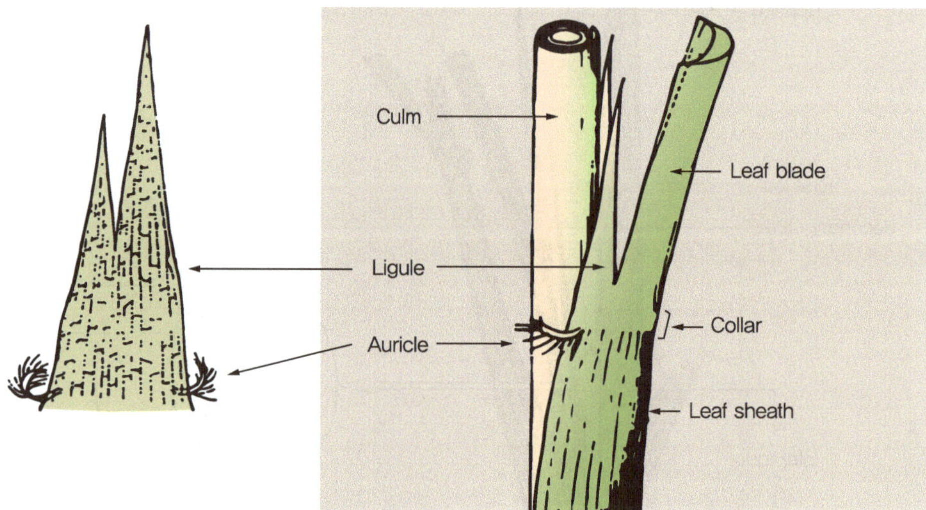

Figure 2-12. Terminology and structure of the rice culm

4) Leaf Structure

(1) Leaf Veins

① The epidermis has a centrally located midrib, with parallel veins extending laterally on both sides.

② Motor cells and stomata are aligned in parallel rows along these veins.

(2) Motor Cells (Bulliform Cells)

① These cells are slightly elongated compared to surrounding cells and are distributed in 2~3 parallel rows between rows of stomata.

② When water is deficient, they contract and cause the leaf to curl inward, reducing transpiration and minimizing water loss.

(3) Stomata

① Stomata are developed on the green epidermis of leaf sheaths, panicles, awns, pedicels, and even hulls (lemma and palea).

② The average size of rice stomata is approximately 30 × 25μm, which is relatively small compared to other plants.

③ The stomatal density ranges from 100 to 600 per mm^2, which is relatively high.

④ The ratio of total stomatal area to total leaf area is about 0.6~0.8%, similar to other plant species.

⑤ Stomatal density increases toward the upper leaves and leaf tips, but decreases under shading treatment.

⑥ *Indica*-type semidwarf varieties have more stomata than temperate *japonica* rice

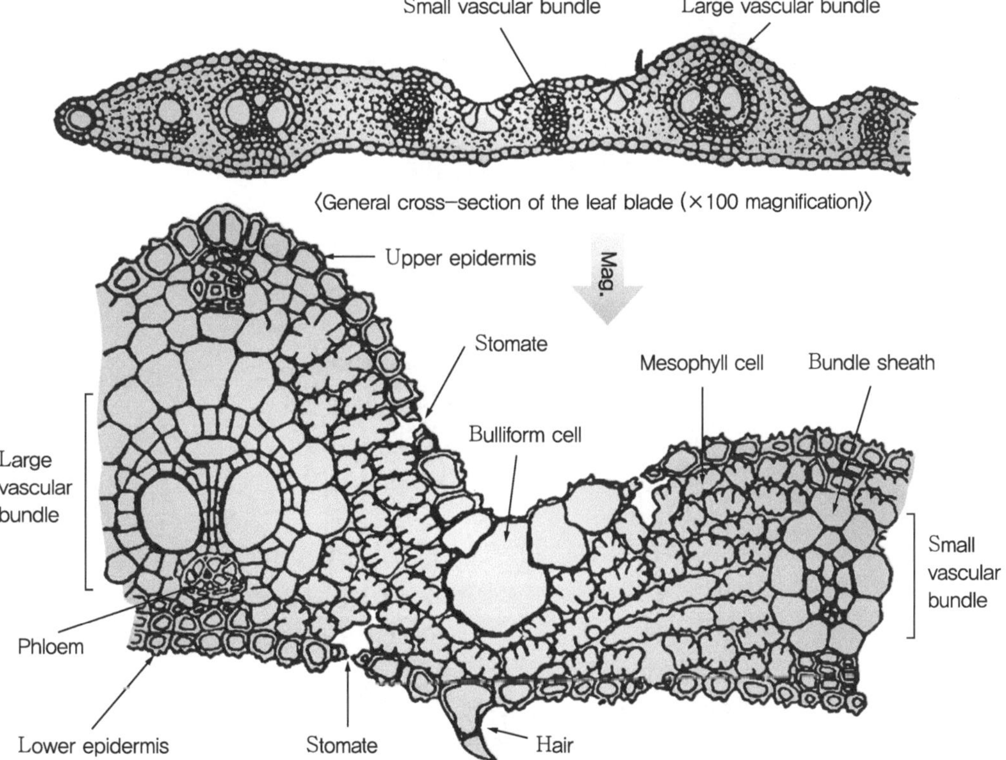

Figure 2-13. Cross-section of the rice leaf blade (Hoshikawa, 1975)

(4) Hydropores

① Hydropores are relatively large stoma-like structures located at the tip of the midrib and vascular bundles in the leaf blade.

② These structures are responsible for guttation, where water droplets appear at the leaf tips, especially in the early morning.

③ Hydropores remain open at all times, and guttation is caused by root pressure.

(5) Internal Structure of the Leaf Blade

① Large vascular bundles develop along the major veins, and small vascular bundles develop along minor veins.

② The midrib is prominently protruded in the center of the leaf blade.

③ Between vascular bundles, mesophyll cells are densely packed in 3 to 5 layers.

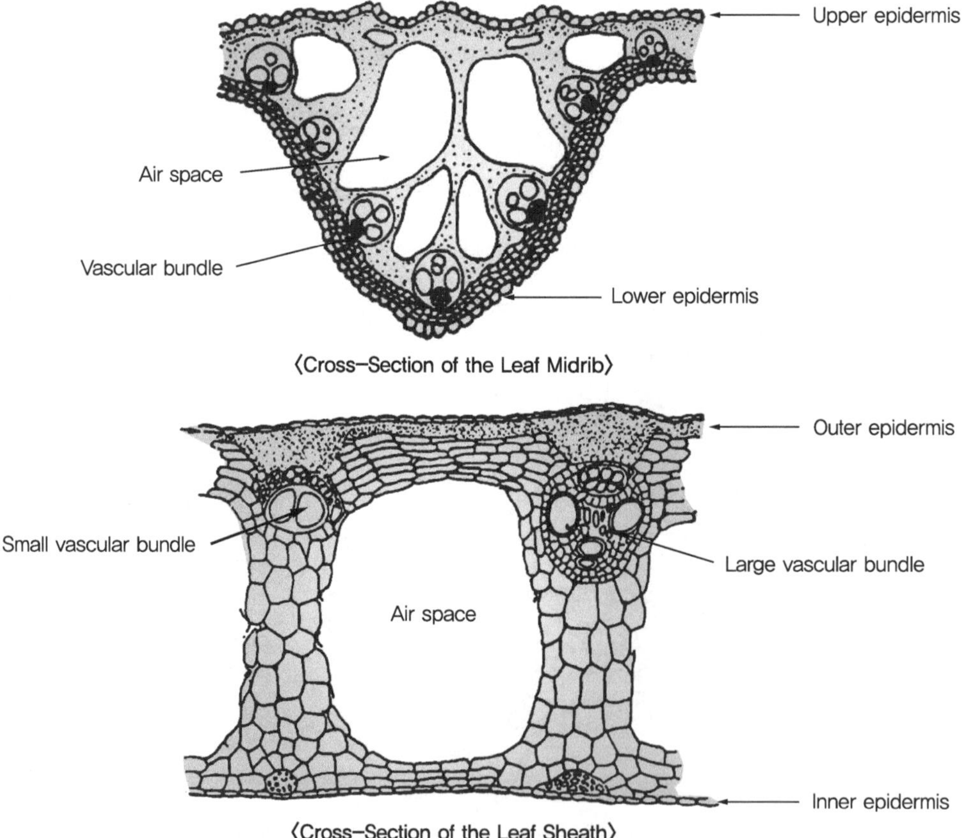

Figure 2-14. Cross-section of the leaf midrib and leaf sheath (Hoshikawa, 1975)

(6) Internal Structure of the Leaf Sheath

① The center is thick and tapers toward the edges. The inner side of the sheath contains air spaces, while the outer side has numerous vascular bundles and sclerenchyma tissues, providing mechanical strength.

② Inside the sheath, a major air cavity called the derived aerenchyma is formed, and one cavity develops for each vascular bundle.

③ These aerenchyma are connected from the stomata to the stem and roots, functioning as oxygen transport channels.

4. Rice Culm (Stem)

1) Morphology of the Stem

① Each node produces one leaf, and therefore the number of nodes equals the number of leaves.

② The internode (the region between two nodes) is cylindrical, encased by the leaf sheath, and has longitudinal grooves on its surface.

③ At each node, a tiller and nodal root are formed, with one leaf and one tiller arising per node.

2) Structure of the Rice Stem

① A rice culm typically has 15 to 20 nodes, with the top five internodes elongating significantly and determining the plant height. The peduncle internode (the internode just below the panicle) is the longest.

② Leaves, tillers, and roots are produced at each node, following a synchronous development theory (coordinated emergence of leaves and tillers).

3) Classification by Culm Position

(1) Basal Culm (Stem Base, Non-elongating Internodes)

① The lower 10~12 nodes of the culm have non-elongated internodes, forming a compact basal region approximately 2cm in length.

② Tiller nodes arise from these lower internodes. As the tiller's point of origin rises (higher position and order), the number of internodes tends to decrease.

② Culm length refers to the distance from the stem base to the panicle base.

(2) Elongating Internodes (Elongation Region)

① During the reproductive growth stage, the upper 5~6 internodes elongate rapidly.

② The number of elongating internodes is similar between the main culm and tiller culms, generally comprising the upper five nodes.

③ The internode length decreases from top to bottom, and the fifth internode from the top is typically only 1~2cm long.

④ The uppermost internode is the longest, reaching around 30cm, and it connects directly to the panicle.

⑤ The neck node (peduncle node) is formed at the first internode below the panicle.

4) Internal Structure of Internodes

① The center of the non-elongating internode contains a large medullary cavity, which is a round, hollow space functioning as a structural column supporting the aerial part of the plant and as an oxygen transport channel.

② The outer layer (epidermis) consists of silicified epidermal cells and several layers of sclerenchyma tissue, and inside it is parenchyma tissue.

③ Numerous small vascular bundles are embedded in the surrounding sclerenchyma, while large vascular bundles are embedded in the internal parenchyma.

④ Each large vascular bundle has a phloem on the outer side and xylem on the inner side and is round in shape.

⑤ Aerenchyma (derived air spaces) develop between large vascular bundles in the parenchyma.

⑥ Aerenchyma are more developed in lowland paddy rice than in upland rice, and more pronounced in floating rice varieties than in standard varieties.

5) Medullary Cavity and Aerenchyma in the Stem

① The hollow medullary cavity observed in cross-sections between nodes acts as a structural support and facilitates oxygen transport.

② The phloem in vascular bundles transports assimilates from leaves to the sink tissues, while the xylem transports water and dissolved minerals from roots upward.

③ Aerenchyma are more developed in the lower internodes of the culm.

5. Rice Floral Organs

The lemma (outer glume) and palea (inner glume) enclose and protect the floral organs. At the base of the lemma and palea are the sterile lemma, rudimentary glume, and the rachilla (floral axis). Rice flowers are perfect flowers, containing both male and female organs, with a Y-shaped pistil and six stamens.

1) Pistil

① The pistil consists of the ovary, style, and stigma. Rice has one Y-shaped stigma, not two separate stigmas.

② The ovary is club-shaped and unilocular, containing a single ovule. The micropyle faces downward, and the ovule has an exposed nucellus and contains an embryo sac internally.

③ The stigma is bifurcated at the tip, feather-like, and colorless to facilitate pollen capture. Depending on the variety, the stigma may appear purple or reddish.

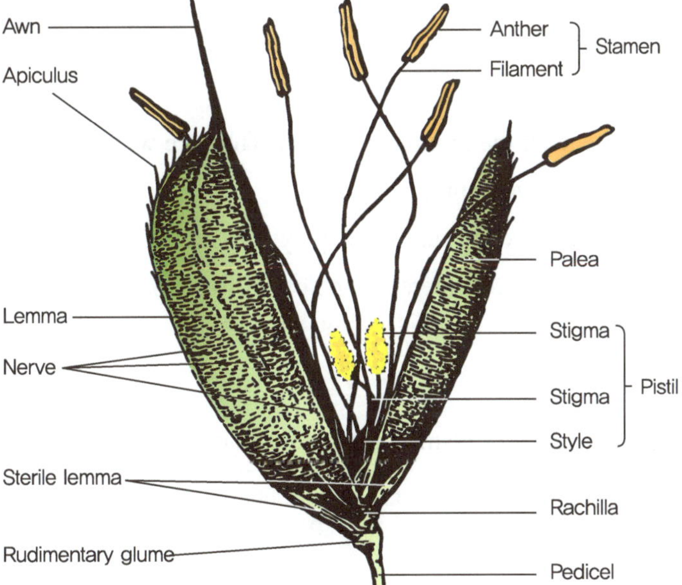

Figure 2-15. Structure of rice floral organs

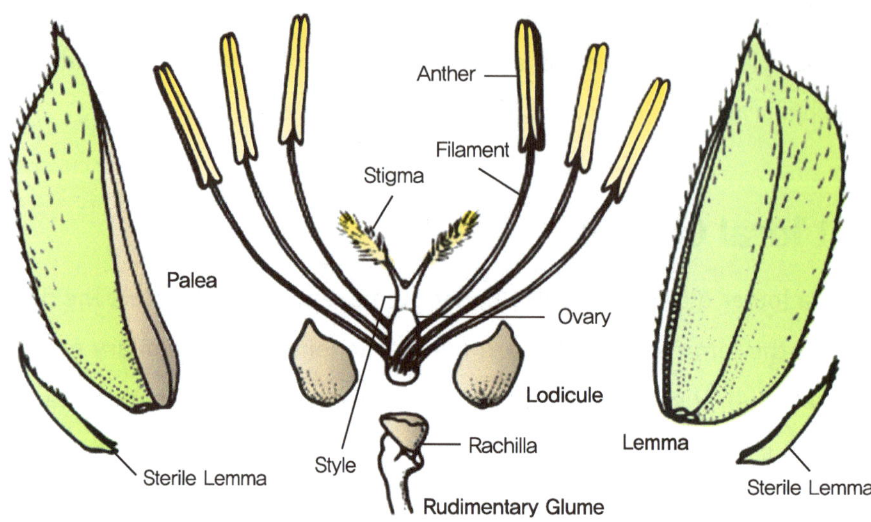

Figure 2-16. Structure and terminology of rice floral organs (Nagai, 1962)

2) Stamen

① The stamen consists of an anther and a filament.

② The anther has four chambers, which contain numerous pollen grains. These pollen grains are spherical in shape and enclosed by a thick outer wall.

③ After flowering, the filament elongates rapidly, causing the anther to burst and release the pollen.

3) Lodicule

① Lodicules are located beneath the lemma and palea. They are white, fleshy, oval-shaped, and usually appear in pairs (two in number).

② During flowering, they absorb water and swell, playing a role in opening the floret.

③ Developmentally, lodicules are considered homologous to petals or tepals.

4) Structure of Rice Flower

Table 2-1. Comparison of the number of pistils and stamens in major crops

Crop	Barley, Wheat, Rye, Oat, Sorghum, Foxtail Millet, Proso Millet	Sweet Potato, Potato	Rice, Lily	Soybean, Peanut, Adzuki Bean, Azalea
Pistil : Stamen ratio	1 : 3	1 : 5	1 : 6	1 : 10

6. Rice Panicle

1) Panicle Structure

① The panicle refers to the part of the rice plant where grains develop, extending upward from the top node of the culm, known as the peduncle node.

② The panicle axis typically contains 8 to 10 nodes, spaced at intervals of about 2~3cm.

③ From each node, a primary branch emerges, followed by secondary, tertiary, and pedicel branches in sequential order.

④ At the end of each pedicel, a spikelet which will develop into a grain is formed. There are usually around 100 spikelets per panicle.

⑤ The number of large vascular bundles in the internode just below the peduncle node is equal to the number of primary branches in the panicle.

⑥ At the center of the panicle neck is a central cavity (cavity in the peduncle), surrounded by 10~12 large vascular bundles arranged in a row, with additional rows of small vascular bundle protrusions aligned along the outer perimeter.

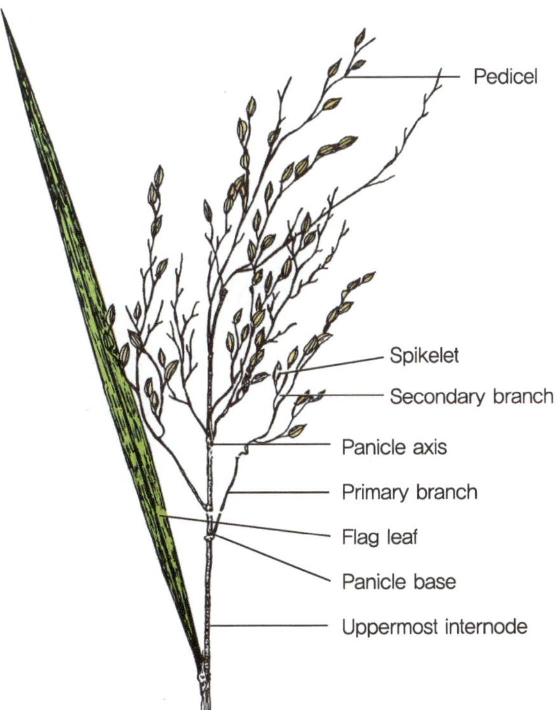

Figure 2-17. Morphological structure of the rice panicle (Chang and Bardens, 1965)

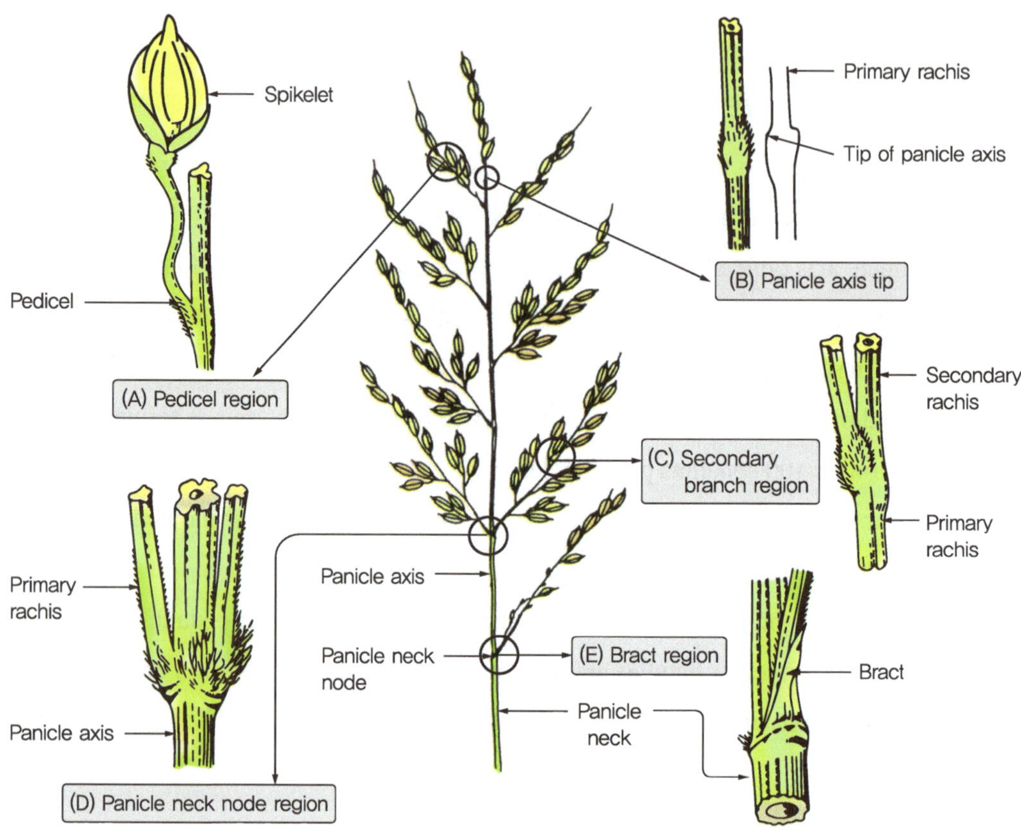

Figure 2-18. Components of the rice panicle (Hoshikawa, 1975)

Rice Crop Science

Chapter 3. **Growth and Development**

Chapter 3.
Growth and Development

The life cycle of rice undergoes a full growing period that includes sowing, germination, growth, flowering, and grain development. The total duration of this period varies depending on the variety and environmental conditions: early-maturing varieties typically require about 120 days, medium-maturing varieties around 150 days, and late-maturing varieties approximately 180 days.

The vegetative growth stage is divided into the nursery stage, transplanting stage, root establishment stage, and tillering stage. During this phase, a large amount of nitrogen is required for protein metabolism. The reproductive growth stage, which includes the heading, flowering, and grain-filling stages, progresses toward flowering and grain formation. The transition between these stages is greatly influenced by atmospheric environmental factors such as temperature and photoperiod, also known as climatic ecological types. Rice growth adapts to the natural environment through metabolic activity. High temperatures shorten the growing period, whereas cooler climates lengthen it.

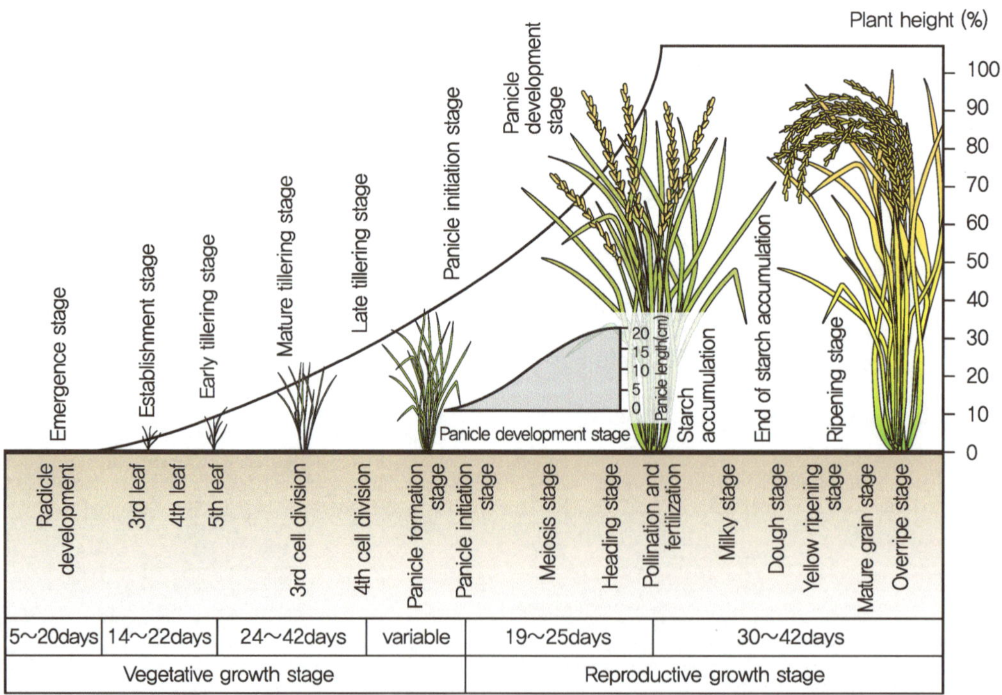

Figure 3-1. Growth stages of rice(Stansel, 1975)

1. Vegetative Growth and Reproductive Growth

1) Vegetative Growth Period

The vegetative growth period refers to the time from germination to just before panicle initiation, during which the vegetative organs leaves, stems, and roots are developed.

The basic vegetative growth period is the minimum period required for developmental maturity and is relatively unaffected by environmental conditions. This juvenile phase is longer in *indica*-type rice, which is mainly cultivated in low-latitude regions, and shorter in temperate *japonica*-type rice, typically grown in higher-latitude regions.

The plastic vegetative growth period is a variable growth phase influenced by environmental factors, particularly temperature and photoperiod. It tends to shorten under high temperature and short-day conditions, and lengthen under low temperature and long-day conditions.

(1) Seedling Raising Stage

① For transplanting cultivation, a seedling raising period is required to grow the young plants.

② "In mechanical transplanting, very young seedlings (10 days), young seedlings (20 days), medium seedlings (30 days), and mature seedlings (40 days, typically used for manual transplanting) are commonly used.

(2) Transplanting Stage & Rooting Stage

① The transplanting stage refers to the time when seedlings are moved from the seedbed to the paddy field. In the case of rice, this process is called "transplanting (이앙)" rather than general "transplantation".

② The rooting stage (also known as the root establishment period) refers to the period after transplanting during which new roots begin to grow.

③ The duration of the rooting period is influenced by factors such as seedling quality, weather conditions, and the appropriateness of transplanting operations.

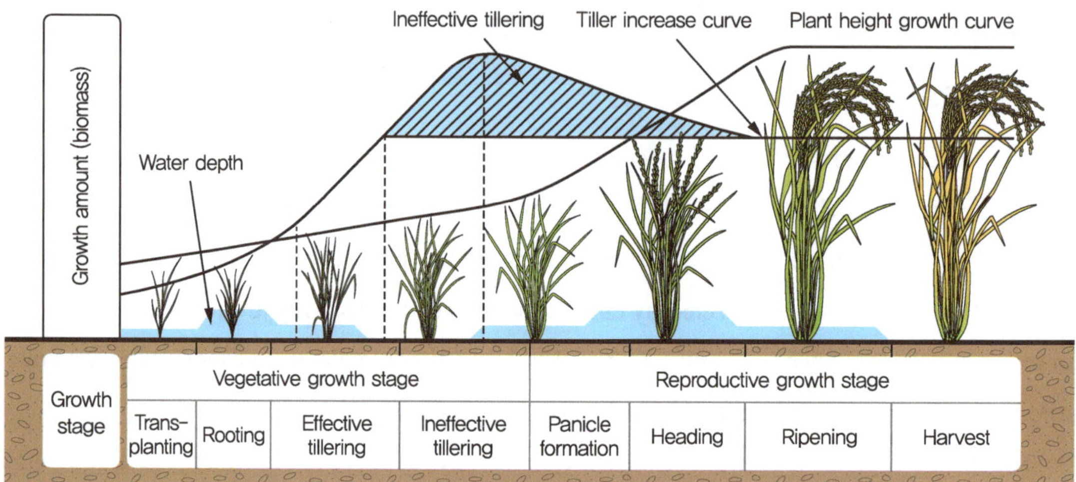

Figure3-2. Growth stages and growth characteristics of rice

(3) Tillering Stage

① The tillering stage is the period during which transplanted seedlings establish roots and begin to produce tillers.

② The tillering burst period is when the number of tillers rapidly increases after transplanting.

③ The end of effective tillering stage marks the transition from producing effective tillers to ineffective ones.

④ The maximum tillering stage refers to the point when the number of tillers reaches its peak after the rapid tillering phase.

⑤ The ineffective tillering stage follows the maximum tillering stage, during which late-emerging, weak tillers die off, resulting in a decrease in total tiller number.

⑥ The effective tiller ratio is the percentage of effective tillers relative to the maximum number of tillers:

(Number of effective tillers / Maximum tiller number) × 100

⑦ Effective tillers are those that develop panicles and successfully produce grain.

⑧ Ineffective tillers are those that fail to form panicles or do not bear grain properly.

⑨ If nitrogen fertilizer is applied excessively or planting density is too high, the number of total tillers increases, but the number of ineffective tillers also rises, lowering the effective tiller ratio.

⑩ The number of days from transplanting to the maximum tillering stage is generally 35 to 40 days under typical cultivation conditions. It becomes longer with earlier transplanting and shorter with late transplanting.

⑪ Generally, tillers that emerge 15 days before the maximum tillering stage and develop three leaves by that time are likely to become effective tillers.

2) Reproductive Growth Period

① The reproductive growth period refers to the phase from panicle initiation to maturity, during which the panicle and floral organs are formed and developed. It includes the period from heading, flowering, and fertilization to ovary development and grain completion.

② When the plant transitions from vegetative to reproductive growth, the upper 4 to 5 internodes of the culm elongate, causing an increase in plant height. At the apical meristem of the stem, cells that would have developed into leaves instead differentiate into panicle cells.

③ From panicle initiation to heading, both vegetative and reproductive growth proceed simultaneously. After heading, only reproductive growth continues. During this aging phase, the number of active leaves and roots decreases, and nutrient uptake declines.

(1) Internode Elongation Stage

① The internode elongation stage is the period from panicle initiation to heading, during which the stem internodes lengthen. This period corresponds to around 30 days before heading, typically near the maximum tillering stage.

② The relationship between panicle initiation and maximum tillering varies by cultivation method: In early planting, panicle initiation occurs after the maximum tillering stage. In normal planting, the two stages coincide. In late planting, panicle initiation occurs before the maximum tillering stage.

(2) Panicle Formation Stage

① The panicle initiation stage begins when the panicle reaches a length of about 2 mm, marking the onset of its differentiation, particularly the development of the panicle axis.

② The floret differentiation stage occurs approximately 7 to 10 days after panicle initiation, when the individual florets begin to differentiate.

③ The panicle formation stage continues until the panicle grows to 3~5cm in length, just before reproductive cells appear inside the anthers approximately 16 days before heading.

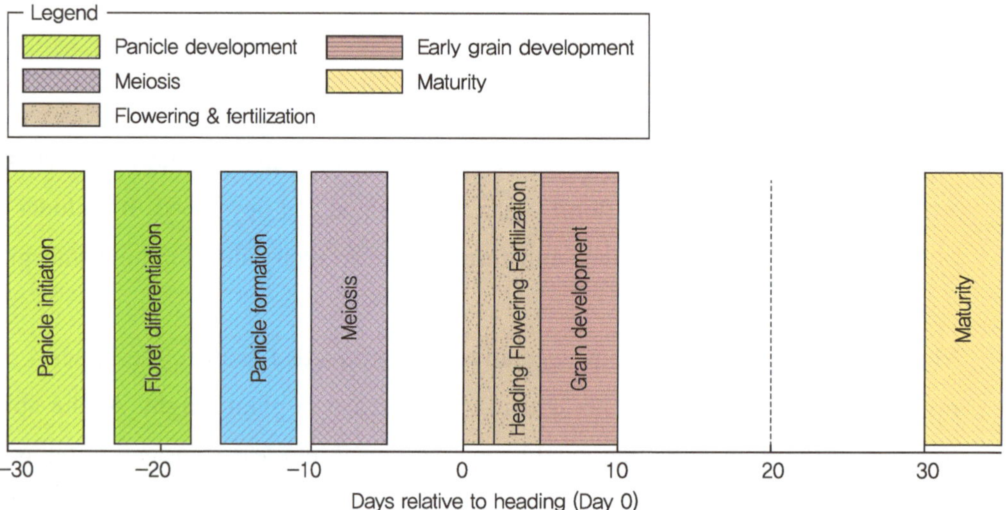

Figure 3-3. Development timeslineof reproductive organs

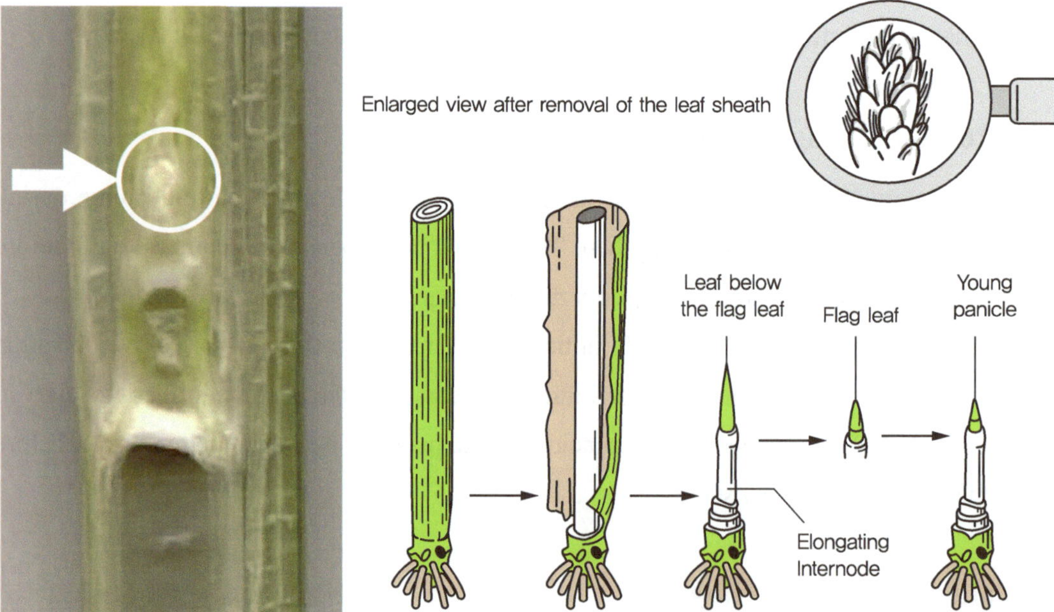

Figure 3-4. Differentiation of the panicle in the rice stem (indicated by arrows), and appearance of the panicle after removal of the leaf sheath

(3) Booting Stage

① The booting stage begins 10~12 days before heading and lasts until just before heading. During this time, the panicle enlarges inside the stem, causing a noticeable swelling.

② During booting, panicle elongation is nearly complete, and pollen mother cells and embryo sac mother cells undergo meiosis, preparing for fertilization.

③ This stage (meiosis stage) is the most sensitive to environmental stress, such as low temperature or drought.

(4) Heading Stage

① The heading stage is when the panicle emerges from the flag leaf sheath. Panicles that emerge typically flower on the same day or the next day.

② In grasses (Poaceae), heading occurs when the peduncle internode elongates, pushing the panicle out of the flag leaf sheath.

③ This stage is considered an important phenological indicator in rice cultivation.

④ The heading stage is defined as follows: when approximately 10% of the panicles have emerged, it is called first heading, when 40~50% have emerged, it is referred to as the heading stage, and when 70~80% have emerged, it is called the full heading stage.

⑤ The duration of heading is shorter when a plant has fewer panicles and longer with more panicles. It is also shorter under high temperatures and longer under low temperatures.

⑥ It takes 2~3 days from first heading to heading stage, and 5~7 days from heading stage to full heading. For one rice plant, full heading may take 7 days, and in a field, it may take 10~14 days.

⑦ If heading occurs when growth is insufficient, it's called premature heading. This often occurs in varieties with high temperature sensitivity, when grown under excessive heat, nitrogen deficiency, or excessively long seedling raising periods.

(5) Ripening Stage

① The ripening stage begins after flowering and fertilization, during which the grain enlarges and matures. Depending on variety and climate, grain maturation takes 30~35 days in warm regions and 45~55 days in cooler regions. High temperatures accelerate ripening, and low temperatures delay it.

② The milky stage is when the grain contents appear milky-white and liquid-like.

③ The dough stage follows, where storage starch increases, moisture decreases, and grains appear gelatinous.

④ The yellow-ripe stage occurs around 30 days after fertilization, when the entire brown rice kernel becomes hard and translucent. Harvesting usually takes place in this stage.

⑤ The mature stage marks full grain maturity, making it suitable for harvest.

⑥ The overripe stage (also called overmature) occurs after optimal harvest time, where the endosperm is completely filled with starch and other storage substances, appearing fully translucent. However, overmature grains are more prone to cracking (chalkiness) during milling.

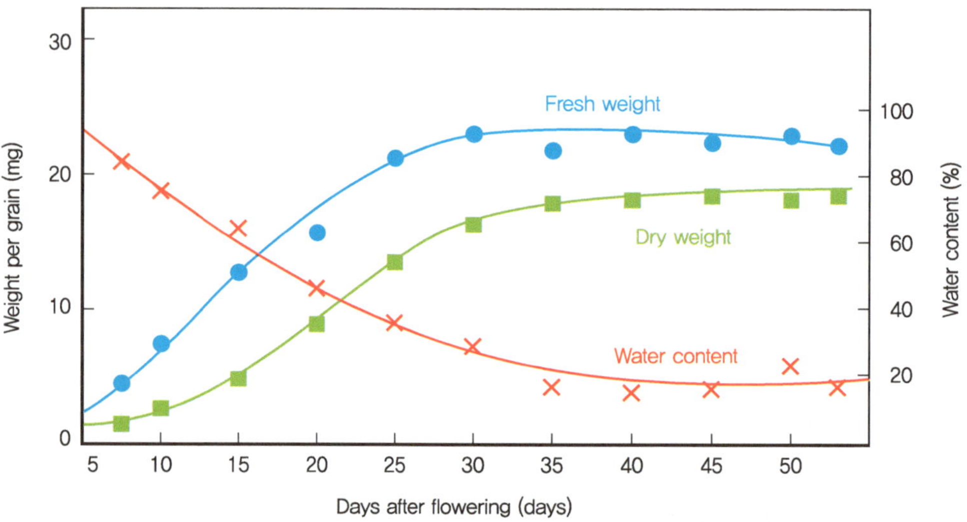

Figure 3-5. Changes in grain weight and water content of brown rice (Hoshikawa, 1975)

2. Germination and Dormancy

1) Seed Longevity

① The decline in seed vigor results from the complex interplay between intrinsic and extrinsic factors of the seed.

② Environmental factors affecting seed longevity include temperature, moisture, light, and atmospheric composition, among which seed moisture content and temperature are the most critical and fundamental.

③ In general, seed longevity increases under conditions of low temperature and low moisture content, particularly when oxygen (O_2) concentration is reduced.

④ In rice (*Oryza sativa*), excessive drying below 15% seed moisture content can paradoxically reduce seed longevity.

2) Seed Dormancy

(1) Dormancy Characteristics

① African cultivated rice (*Oryza glaberrima*) and wild rice species exhibit strong seed dormancy.

② The *indica* rice subspecies tends to have stronger dormancy than the temperate *japonica* types.

③ Tongil-type varieties (Tongilbyeo, a hybrid between *indica* and *japonica*) generally exhibit moderate dormancy, stronger than typical temperate *japonica* varieties.

④ In cultivars with weak dormancy such as temperate *japonica* rice, pre-harvest sprouting (PHS) can easily occur under high humidity or prolonged rainfall during harvest time.

⑤ Immediately after harvest, blastokoline, a germination-inhibiting compound present in the lemma and palea (hull), suppresses germination and induces dormancy.

(2) Dormancy Breaking

① In rice, removal of the hull, which contains germination inhibitors, can break dormancy. Depending on the genotype, partial or complete removal of the pericarp and seed coat may also be required to induce germination.

② Exposing rice seeds to high temperature (50°C) for 4~5 days, or 7~10 days for deeply dormant varieties, effectively breaks dormancy.

③ Treatment with nitric acid (HNO_3), such as soaking 1.2kg of dry seeds in 1L of 0.1N HNO_3 for 1~2 days (or 3~7 days for stronger dormancy), followed by drying, can also break dormancy in rice seeds.

3) Germination

Germination refers to the process in which the embryo resumes growth after dormancy by absorbing water, resulting in a visible radicle elongation of about 1mm.

In contrast, seedling emergence refers to the stage where the coleoptile or shoot protrudes above the soil surface.

(1) Movement of Absorbed Water

① Water absorption in rice seeds occurs primarily through the scutellum, the interface between the embryo and endosperm. Upon imbibition, hydrolytic enzymes are synthesized in the aleurone layer and mobilized toward the endosperm.

② The absorbed water moves into the embryo tissues via the absorptive cell layer of the scutellum.

③ Water also travels along the aleurone layer toward the micropylar end of the seed.

(2) Germination Process

① Once water is absorbed, the aleurone layer synthesizes hydrolytic enzymes (e.g., amylases, lipases, proteases) and transfers them into the starchy endosperm.

② These enzymes hydrolyze macromolecular storage compounds, starch, lipids, and proteins, into simple molecules such as sugars, fatty acids, and amino acids.

③ The resulting metabolites are translocated into the embryo, where they are utilized for the growth of pre-differentiated cells in the plumule and radicle.

④ As the seed enters the germinative state, its respiration rate increases sharply, indicating the onset of active metabolic processes.

4) Factors Influencing Germination

(1) Conditions for Seed Germination

① Although rice seeds can germinate as early as 7 days post-fertilization, immature seeds require more days to germinate and exhibit incomplete development.

② Around 14 days after fertilization, both germination rate and germination speed become nearly normal.

③ Among seeds of the same cultivar, heavier seeds (higher specific gravity) tend to exhibit greater vigor and better post-germination growth.

④ Seed vigor also varies within the same panicle: upper spikelets generally show faster germination and higher viability than lower spikelets.

⑤ Prolonged storage duration reduces germination rate, and under natural conditions, germinability declines sharply after two years.

(2) Water Availability

A: Imbibition phase

① For germination to occur, rice seeds must absorb approximately 30–35% of water based on dry weight, and germination can begin when about 23% of the seed's weight is water.

② The initial imbibition phase (Phase A) is a passive and physical water absorption stage.

③ During imbibition, water uptake initiates physiological activation in both the embryo and endosperm.

④ This phase proceeds rapidly until a threshold moisture content is reached, with temperature exerting minimal influence.

⑤ The seed typically reaches the required moisture content for germination within about 18 hours.

⑥ The embryo becomes physiologically active when the seed moisture reaches approximately 15% of seed weight.

⑦ Higher temperatures accelerate the water absorption rate during imbibition.

B: Activation phase (Preparatory phase for germination)

① This is the metabolic activation period following sufficient water uptake.

② In this phase, seeds maintain a moisture content of 30`35%, initiating germination-related physiological and biochemical processes.

③ It is highly temperature-sensitive, encompassing enzyme activation in the early stage, followed by enzymatic hydrolysis of endosperm reserves and their translocation into the embryo in the later stage.

④ Translocated sugars are partially consumed via cellular respiration and partially stored as energy reserves to support plumule and radicle growth.

⑤ By the end of this phase, the embryo produces a visible sprout, marking the initiation of germination.

C: Growth phase (Post-germination)

① This phase begins when the plumule and radicle break through the seed coat and elongation occurs via cell expansion.

② Water uptake continues and increases rapidly, supporting cellular metabolism and growth.

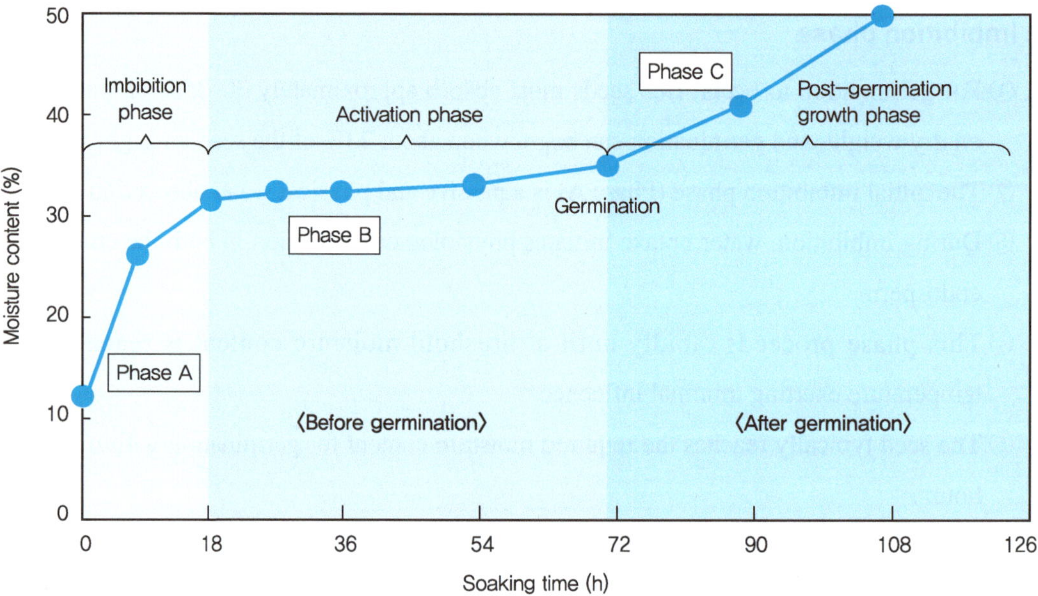

Figure 3-6. Water uptake pattern during rice seed germination (Takahashi, 1961)

(3) Temperature

① The optimum, minimum, and maximum temperatures for rice seed germination vary depending on the ecotype or cultivar.

② Generally, the minimum germination temperature is 8~10°C, the optimum is 30~32°C, and the maximum is 44°C.

③ There is a wide variation in minimum germination temperature among cultivars. Cold-tolerant varieties from high-latitude regions exhibit stronger low-temperature germination ability compared to tropical varieties from low-latitude regions.

④ Traditional Korean landraces can germinate even at 8°C, whereas some tropical rice seeds may not germinate at temperatures as high as 16~24°C.

⑤ Rice seeds with high vigor germinate within 24~48 hours after sowing at 30°C.

⑥ When seed dormancy is fully broken and vigor is high, varietal differences in germination rate are minimal. However, seeds with insufficient dormancy release or reduced vigor exhibit a narrower germination temperature range.

Table 3-1. Temperature requirements for each growth stage of rice (Yoshida, 1977)

No.	Growth stage	Minimum temp. (°C)	Optimum temp. (°C)	Maximum temp. (°C)	Remarks
①	Germination	10	20–35	45	Germination occurs across a wide temperature range.
②	Emergence (seedling stage)	12–13	25–30	35	-
③	Rooting (establishment)	16	25–28	35	-
④	Vegetative growth	7–12	31	45	Vegetative and tillering stages tolerate relatively low temperatures.
⑤	Tillering	9–16	25–31	33	
⑥	Panicle initiation	15–20	-	38	Panicle initiation and heading stages require higher minimum temperatures.
⑦	Heading (flowering)	22	30–33	35	
⑧	Maturation	12–18	20–25	30	The maturation stage typically occurs under declining temperatures.

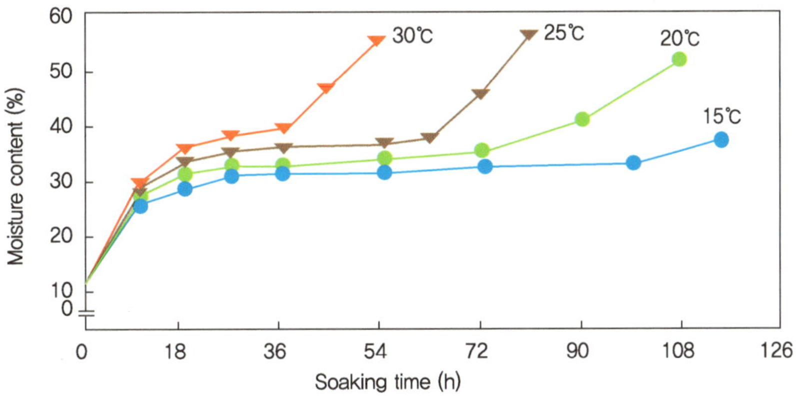

Figure 3-7. Water absorption of rice seeds under different germination temperatures (Takahashi, 1961)

(4) Oxygen (O_2)

A: Aerobic conditions

① Rice seeds are capable of germinating both in water and in air, having adapted to conditions of both flooded paddy fields and dry upland soils.

② Under conditions with sufficient oxygen supply, the radicle (primary root) emerges first and grows normally.

③ In oxygen-rich environments, the coleoptile emerges as a short (<1cm), thick structure, accompanied by the growth of seminal roots.

④ When the germinated seedling (coleoptile) grows above the water surface in deep water, the supply of oxygen promotes further radicle growth.

⑤ Since root anchorage is difficult under submerged conditions, drainage is necessary during the rooting stage to facilitate oxygen supply.

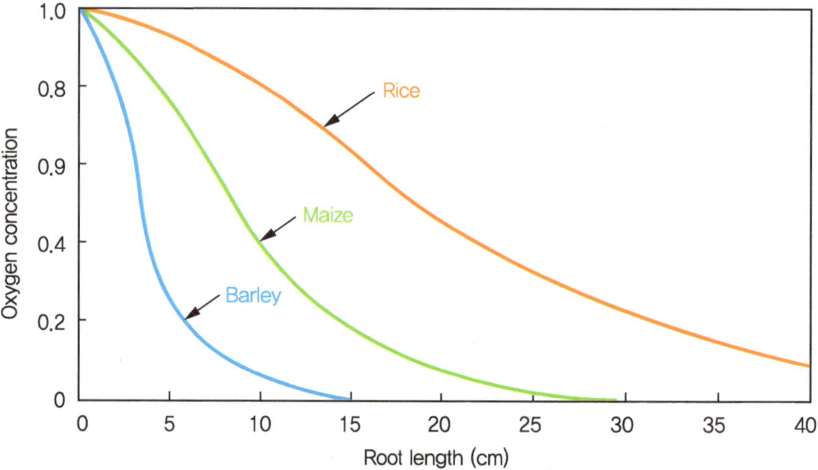

Figure 3-8. Effect of oxygen concentration on root length (Jensen et al., 1967)

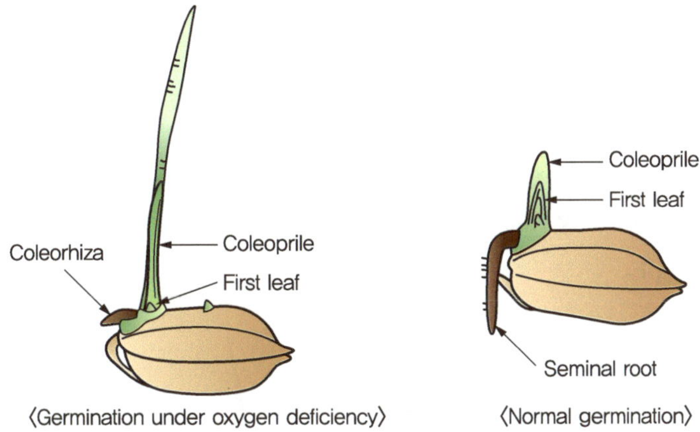

Figure 3-9. Germination patterns of rice seeds under different oxygen conditions

B: Anaerobic conditions

① Under oxygen-deficient conditions, the activity of enzymes essential for germination, such as catalase, cytochrome oxidase, and amylase, is significantly reduced.

② Rice seeds require less oxygen for germination compared to other crops and can achieve a 100% germination rate even at an oxygen concentration of approximately 0.5%.

③ Even under anoxic (oxygen-free) conditions, rice seeds exhibit about 80% germination via anaerobic respiration. In oxygen-deficient and dark environments, the mesocotyl elongates considerably.

④ When oxygen is limited, the coleoptile may elongate, but the first leaf and radicle show poor development.

⑤ Rice seedlings germinated under oxygen-deficient, dark conditions may exhibit coleoptile elongation up to 4~6cm, with extensive mesocotyl elongation, resulting in abnormal morphology.

⑥ When oxygen is insufficient, radicle growth is suppressed while the embryo axis (seedling) elongates first.

(5) Light Intensity

① Rice seeds do not require light for germination and can germinate in complete darkness, but seedling (etiolated organ) growth after germination is affected by light availability.

② The mesocotyl refers to the internode between the coleoptilar node and the coleorhizal node (the region between the coleoptile base and the seminal root's coleorhiza) and generally does not elongate under normal conditions.

③ Under dark conditions, the mesocotyl elongates, producing a phenotype similar to that observed under low oxygen conditions.

④ The mesocotyl plays a key role in pushing the coleoptile above the soil surface when seeds are sown deeply.

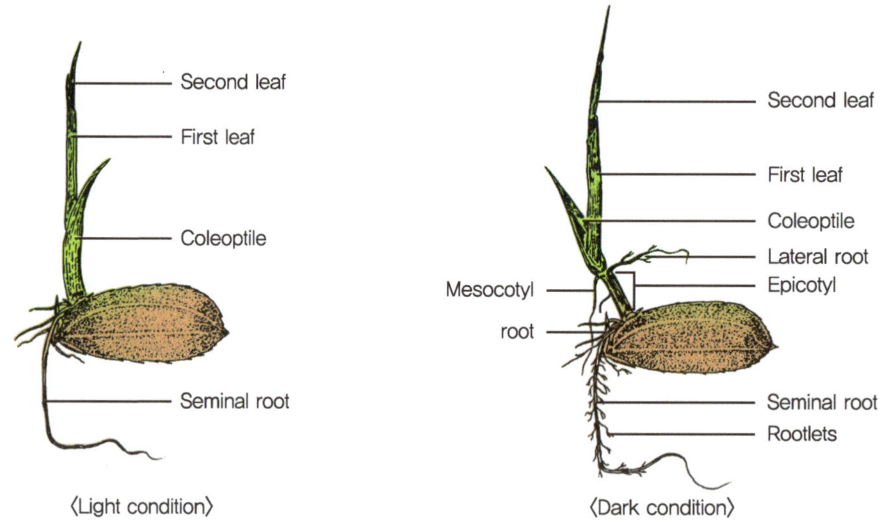

Figure 3-10. Germination morphology of rice under light and dark conditions (Chang and Bardens, 1965)

(6) Seeding Depth

① When seeds are sown at a shallow depth of 0.5cm near the soil surface, oxygen availability is high and light penetration is sufficient, resulting in favorable germination and seedling growth.

② When sown at a depth of 2cm below the soil surface, some etiolation (excessive elongation of seedlings) may occur.

③ At a seeding depth of 3cm, the mesocotyl and coleoptile elongate excessively, making emergence above the soil surface more difficult.

④ Even in dry direct seeding conditions, a depth of 3cm is considered the practical limit.

⑤ In paddy soils, sowing at a depth of 3cm causes elongation of the mesocotyl and the emergence of crown roots at the coleoptile node, and at 5cm depth, mesocotyl roots develop and grow horizontally.

3. Growth of Rice Seedlings

1) Growth of Seedlings

① A germinated seed initially develops a single seminal root, and around the weaning stage, 5 to 6 crown roots begin to emerge.

② Approximately 20 days after sowing, the seminal root reaches its full length and the number of crown roots increases rapidly.

③ In hand-transplanted seedlings, a mature seedling form is established around 40 days after sowing.

④ The nitrogen content of the seedling peaks at the 4th to 5th leaf stage, after which it declines and the C/N ratio increases.

2) Weaning Stage of Seedlings

① Until the emergence of the third true leaf, the seedling relies on the nutrients stored in the endosperm for growth (heterotrophic growth).

② The stage at which the seedling transitions from heterotrophic to autotrophic growth, generally around the 3rd to 4th leaf stage, is referred to as the weaning stage.

③ After the 4th leaf stage, growth is supported by nutrients absorbed through the newly developed roots.

④ There are three periods when the leaf color of seedlings tends to turn yellow: the emergence of the first true leaf from the coleoptile, the depletion of endosperm reserves (at the appearance of the 4th leaf, marking the shift from heterotrophic to autotrophic growth), and the late nursery period.

⑤ If the seedlings are sown too densely or too late, or if nutrients are deficient, especially under low light conditions, etiolation may occur, increasing susceptibility to pests and diseases. Therefore, careful nursery management is essential.

4. Root Growth

1) Root System Formation

(1) Root Development

① Root development in rice is closely associated with tiller and leaf emergence. According to the "simultaneous leaf-tiller-root theory," the nodal root and tiller at node n are formed concurrently with the emergence of the leaf at node n+3.

② The ratio of root dry weight to shoot dry weight is about 35% in the early growth stage, 16% at heading, and decreases to about 7% during grain filling. This indicates a gradual decline in root proportion as growth progresses.

③ In direct-seeded rice, roots are more concentrated near the soil surface compared to transplanted rice. In dry direct seeding, more roots initially grow deeper into the soil than in water-seeded rice.

④ To achieve high yields, it is essential to develop and maintain a large number of vigorous roots with high activity until maturity.

(2) Root System

① Approximately 45% of the total roots are distributed within 5cm of the soil surface, and about 90% within 20cm.

② During the tillering stage, roots spread approximately 35cm horizontally and 20cm vertically.

③ At heading, roots may extend to around 40cm horizontally and 50cm vertically, with maximum depths reaching up to 90cm.

④ In the early growth stages, roots are shallow and concentrated near the surface, forming a flat oval pattern. As tillering progresses, roots expand laterally and penetrate deeper soil layers.

(3) Number and Length of Roots

① The number and length of roots begin to increase after transplanting and peak around the heading stage.

② The total root length, including both primary and lateral roots, continues to increase even after heading, but root activity begins to decline thereafter.

③ The maximum number of tillers usually occurs about 35 days before heading, just before panicle initiation.

④ The highest number of roots is observed around 15 days after the maximum tillering stage or about 20 days before heading, while the longest root length is typically achieved at heading.

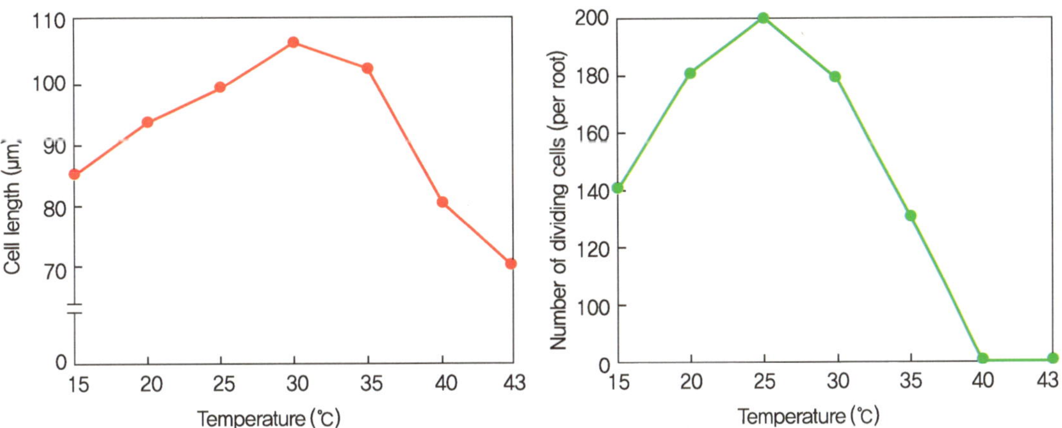

Figure 3-11. Cell elongation and cell division in the radicle as affected by temperature (Yamakawa and Kishikawa, 1957)

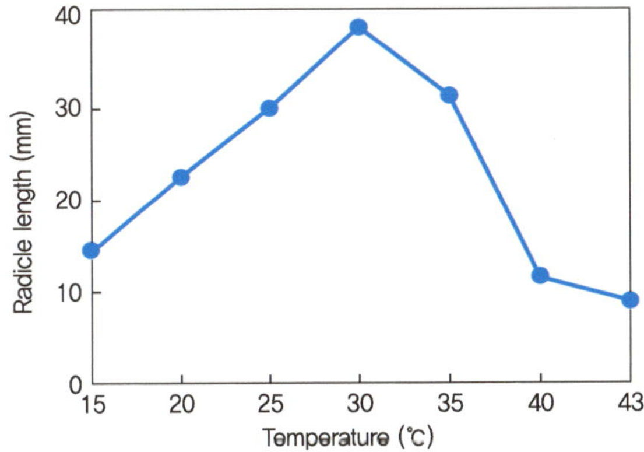

Figure 3-12. Radicle elongation in relation to temperature (Yamakawa and Kishikawa, 1957)

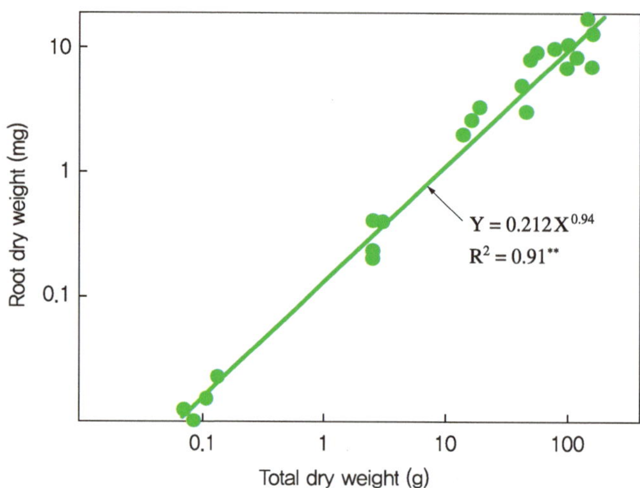

Figure 3-13. Relation between total dry weight (X) and root dry weight (Y) in four rice cultivars (IRRI, 1978)

Table 3-2. Classification of rice roots (IRRI, 1981)

Root group	Root hair (%)	Average length (cm)	Average age (days)
Group 1	None	5~7	≤3
Group 2	~50%	10~20	3~7
Group 3	~80%	15~30	7~14
Group 4	≥90%	≥30	≥14

Table 3-3. Comparison of characteristics between tillering- and panicle weight-type rice

No.	Trait	Tillering-type	Heavy panicle
①	Stem diameter	Thin	Thick
②	Tillering ability	High	Low
③	Number of primary roots	Many	Few
④	Root type	Shallow-rooted	Deep-rooted
⑤	Root size	Small	Large

2) Environmental Factors Affecting Root Growth in Rice

(1) Soil Conditions

① In aerobic upland soil conditions (compared to flooded paddy fields), the development of root hairs is more active due to higher oxygen availability.

② In submerged paddy soils with low oxygen (O_2), root elongation and the formation of lateral roots are reduced, resulting in thinner roots.

③ In upland soils with sufficient oxygen, crown roots elongate and produce many lateral roots.

④ Under reduced (anaerobic) conditions in flooded soils, rice roots develop aerenchyma in the cortical tissue, a specialized adaptation to transport oxygen from the shoots to the roots.

⑤ The proportion of aerenchyma in the root cortex increases from about 25% under upland conditions to up to 60% under flooded conditions.

⑥ Oxygen transported from the shoot is released at the root apex, oxidizing the rhizosphere and allowing root penetration into reduced soil layers.

⑦ The released oxygen combines with iron in the soil to form reddish-brown iron oxide (Fe^{3+}) coatings, which protect the roots from toxic gases such as hydrogen sulfide (H_2S) and methane (CH_4) generated under anaerobic conditions.

(2) Water Management Conditions

① In soils with high percolation rates, a greater proportion of primary roots grow vertically into the soil.

② Compared to continuously flooded soils, well-drained or intermittently irrigated soils promote an increased number and greater length of primary roots.

③ Since flooded soils often become oxygen-deficient in the root zone, midseason drainage or intermittent irrigation during vigorous root growth is recommended to enhance oxygen supply.

④ Paddy fields with strong soil reduction require drainage systems to promote aerobic soil conditions.

(3) Oxygen Availability

① Adequate oxygen enhances root elongation and nutrient uptake via efficient respiratory energy production, which also stimulates shoot growth.

② Oxygen deficiency inhibits root growth due to accumulation of harmful gases such as H_2S and CH_4, which damage root tissues.

③ When iron (Fe) is sufficiently available, rice forms protective iron oxide coatings on the root surface to prevent H_2S damage.

④ Rice is well adapted to submerged conditions due to well-developed aerenchyma in its root cortex.

⑤ Rice roots release oxygen from the tips, creating oxidized zones in the surrounding soil (oxidation rhizosphere effect).

(4) Fertilization Conditions

① Nitrogen (N) fertilizer has a particularly significant effect on root system development in rice.

② Excessive nitrogen application reduces the overall root system size, increases the number of primary roots, but shortens their length, with roots more concentrated in the upper soil layer.

③ When a large portion of nitrogen is applied as basal fertilizer, fewer surface roots are formed. In contrast, frequent split applications promote surface root proliferation.

④ Deep placement of nitrogen results in greater development of deeper-penetrating primary roots compared to surface application.

(5) Planting Conditions

① Increasing the number of seedlings per hill increases the number of primary roots per plant, but reduces root diameter and the number of deep-penetrating roots.

② Higher planting density decreases the proportion of primary roots that grow deeply into the soil.

5. Leaf Growth in Rice

1) Leaf Initiation and Development

① Leaf primordia are sequentially differentiated and developed from the shoot apical meristem.

② Both the leaf blade (lamina) and leaf sheath begin to elongate from the tip and gradually expand toward the base.

③ During the development of a single leaf, elongation begins first in the blade, and once it has sufficiently elongated, the sheath starts to grow.

④ As the sheath of the nth leaf elongates, the blade of the (n+1)th leaf also begins to extend so that both grow simultaneously.

⑤ Rapid leaf expansion and maintenance of an optimal number of functional leaves are crucial for sustaining high photosynthetic capacity and supporting rice growth.

2) Leaf Emergence (Leaf Appearance Rate)

① High temperatures accelerate leaf emergence, whereas low temperatures delay it.

② One leaf emerges every 4~5 days before panicle initiation and every 7~8 days after panicle initiation.

3) Leaf Longevity

① The lifespan of a leaf refers to the duration during which it performs photosynthesis after full expansion.

② The duration of leaf activity is shorter in lower leaves and longer in upper leaves, with the flag leaf (the uppermost leaf) exhibiting the longest activity period.

③ Leaves that have completed their function senesce starting from the basal older leaves.

④ Between panicle initiation and heading, approximately five actively photosynthesizing leaves are present on the culm.

⑤ Under low light conditions, leaves become longer and thinner, and their longevity is reduced.

4) Functionally Dominant Leaf

① The functionally dominant leaf refers to the leaf with the highest photosynthetic activity at a given growth stage.

② During the vegetative stage, photosynthetic activity is higher in the third to fourth leaf below the topmost developing leaf (not in the topmost itself).

③ As lower leaves age and senesce, newly developed leaves take over, and the functional center shifts upward.

④ In the reproductive stage, the flag leaf and the one just below it become the main photosynthetically active leaves (functional center leaves).

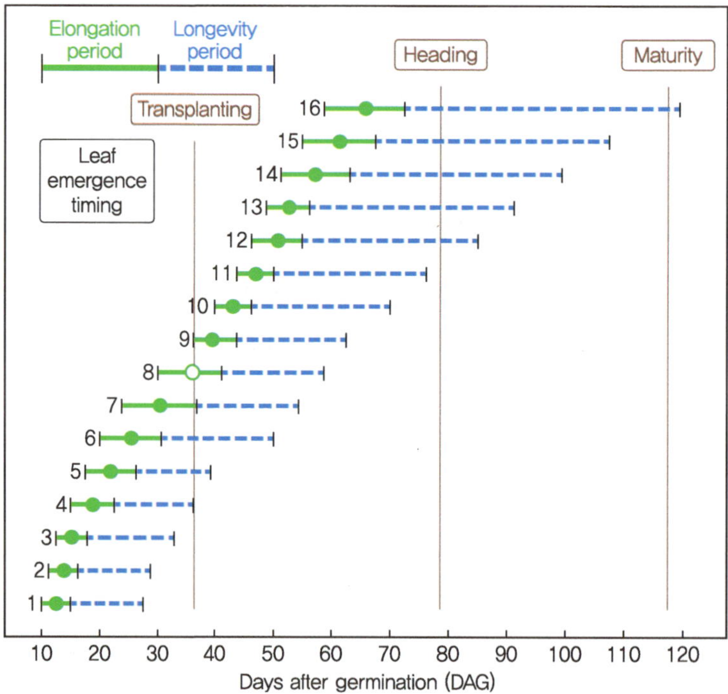

Figure 3-14. Leaf blade elongation and longevity by leaf position in rice (Arashi, 1955)

5) Leaves and Translocation

① The direction of translocation of photosynthates is influenced by the sink demand and the spatial proximity between the leaf and the sink organ.

② Grain filling in rice mainly depends on the assimilates produced by the upper leaves, whereas root activity is maintained by the lower leaves.

③ Assimilates synthesized in the uppermost flag leaf are preferentially translocated to the developing panicle.

④ Assimilates from the lowermost leaves are transported to nearby roots, where demand is high.

6) Leaf Sheath

① Until the panicle initiation stage, when internode elongation is minimal, the leaf sheath functions as a mechanical support structure for the rice plant.

② After heading, once internode elongation is complete, the sheath continues to support the culm and prevent lodging. It also translocates stored carbohydrates (accumulated before heading) to the panicle, contributing to grain yield.

③ Leaf biomass rapidly increases during leaf elongation and reaches a peak, then gradually decreases as assimilates are mobilized to developing upper leaves.

7) Leaf Growth and Functional Activity

① Leaf growth in rice is expressed in terms of leaf area per unit ground area, leaf dry weight, and leaf area index (LAI).

② The LAI increases during vegetative growth, reaches a maximum around heading, and gradually declines thereafter.

③ The most influential agronomic factors affecting total leaf area are planting density and nitrogen application rate, as they determine leaf size, number of leaves per culm, and number of tillers per plant.

④ During the vegetative phase, the 3rd to 4th uppermost leaves are the main photosynthetically active leaves, and during the reproductive phase, the flag leaf and the second uppermost leaf act as the physiological center of photosynthetic activity.

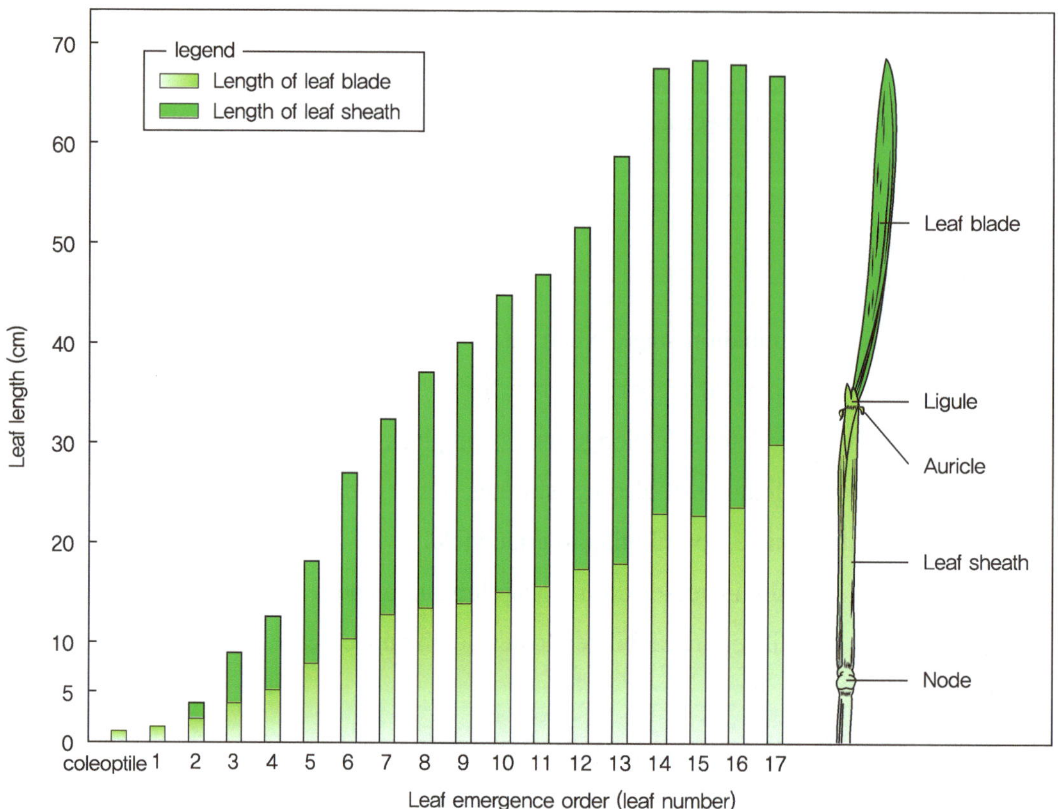

Figure 3-15. Length of leaf sheath and leaf blade by leaf position in rice (Hoshikawa, 1975)

6. Culm (Stem) Growth in Rice

1) Internode Elongation

① The stem directly emerging from the seed is referred to as the main culm, and it consists of nodes and internodes.

② The number of nodes on the main culm corresponds to the number of developed leaves.

③ When vegetative growth is suppressed, the number of nodes decreases, and if panicle initiation occurs prematurely, 4 to 6 internodes may be reduced.

④ Internode elongation begins approximately 35 days before heading and reaches its maximum at the heading stage.

2) Tillering

(1) Tillering System

① A tiller refers to a lateral shoot that arises from the axillary bud at the node.

② After transplanting and the establishment phase, plant height increases and tiller production becomes vigorous.

③ Tillers on the main culm do not arise from the coleoptile node or the first leaf node, but rather from the unelongated basal nodes starting at the second leaf node.

④ Before the emergence of the first true leaf, a prophyll (initial scale leaf) appears as a precursor during tiller formation.

(2) Synchronous Growth

① Synchronous tillering occurs in coordination with leaf and root development in a regular sequence.

② The first leaf of the tiller originating from the axils of the nth and (n–3)th leaves grows simultaneously.

③ When the fifth leaf of the main culm emerges, a tiller and adventitious roots simultaneously appear at the second node.

④ This synchronized pattern applies consistently to secondary and tertiary tillers as well.

(3) Environmental Factors Affecting Tillering

A: Temperature

① The basal internode (culm base) is more sensitive to temperature than the root zone.

② The optimum air temperature for tillering is approximately 26°C, and the optimum water temperature is around 30°C, indicating that water temperature has a greater influence than air temperature.

③ Tillering is most active when daytime temperatures are 30~35°C and nighttime temperatures are around 15°C, with a day–night temperature difference of 10~15°C being ideal.

④ Although the optimal tillering temperature range is 20~25°C, a larger day–night thermal amplitude within this range promotes increased tiller production.

⑤ In early-season cultivation, which often occurs under relatively cool conditions with large day–night temperature differences, the number of tillers tends to be higher than in normal-season planting.

B: Light intensity

① Higher light intensity increases tiller number, particularly during the early tillering phase.

② When the average solar radiation is below 200 $cal/cm^2/day$, tiller emergence is significantly inhibited.

C: Water availability

① In rice, overall growth, including tillering, is suppressed under water-deficient soil conditions.

② Deep flooding reduces temperature and narrows the day–night temperature range, which inhibits tillering.

③ Tillering can be suppressed at a water depth of 5cm during the seedling stage, whereas during peak tillering, depths of up to 10cm have little effect.

D: Nutrient availability

① Adequate supply of inorganic nutrients and photosynthates is essential for tiller formation and development.

② To promote vigorous tillering, the active leaves should contain at least 3.5% nitrogen (N) and 0.25% phosphorus (P_2O_5).

③ During the tillering stage, leaf nitrogen content should be ≥2% to ensure optimal photosynthesis.

④ Panicle nitrogen topdressing is most effective when leaf nitrogen content is ≤2%, and increased 1,000-grain weight is observed when nitrogen exceeds 1.2%.

E: Seedling age

① Under favorable conditions, tillering begins from the second node.

② In transplanted rice, seedlings grown in densely sown nursery beds experience dormancy in lower node buds due to overcrowding, so tillers do not form from the lower nodes even after transplanting.

③ Tillers arise from the fifth node in 40-day-old manually transplanted seedlings, from the fourth to fifth node in 30-day-old mechanically transplanted seedlings, and from the fourth node in 20-day-old machine-transplanted seedlings.

F: Transplanting depth and planting density

① When seedlings are transplanted too deep (deep transplanting), the soil temperature decreases and the thermal difference narrows, leading to delayed root establishment and suppressed tillering. Primary tillers then arise from higher nodes, reducing the number of productive tillers.

② Increased planting density leads to overcrowding, which reduces the number of tillers per individual plant.

③ When the number of seedlings per unit area is the same, planting more hills with fewer seedlings per hill promotes tiller emergence from lower nodes and increases the number of tillers and panicles.

④ Deep planting of densely sown seedlings causes dormancy of tiller buds at the lower 3~4 nodes, reducing overall tiller production.

G: Direct seeding cultivation

① In direct-seeded rice, tillering occurs from the 2nd to 12th nodes, producing more tillers than transplanted rice, but also increasing the proportion of non-productive tillers.

② In contrast, transplanted rice typically shows tillering between the 5th and 10th nodes.

③ Since direct-seeded rice develops a shallow root system but excessive tillering, lodging (culm bending) becomes more frequent.

(4) Effective vs. Ineffective Tillers

A: Effective tillers

① An effective tiller is one that develops a panicle and produces grains.

② A high proportion of effective tillers indicates fewer non-productive tillers, reduced energy waste, and is a key indicator for achieving high yield potential.

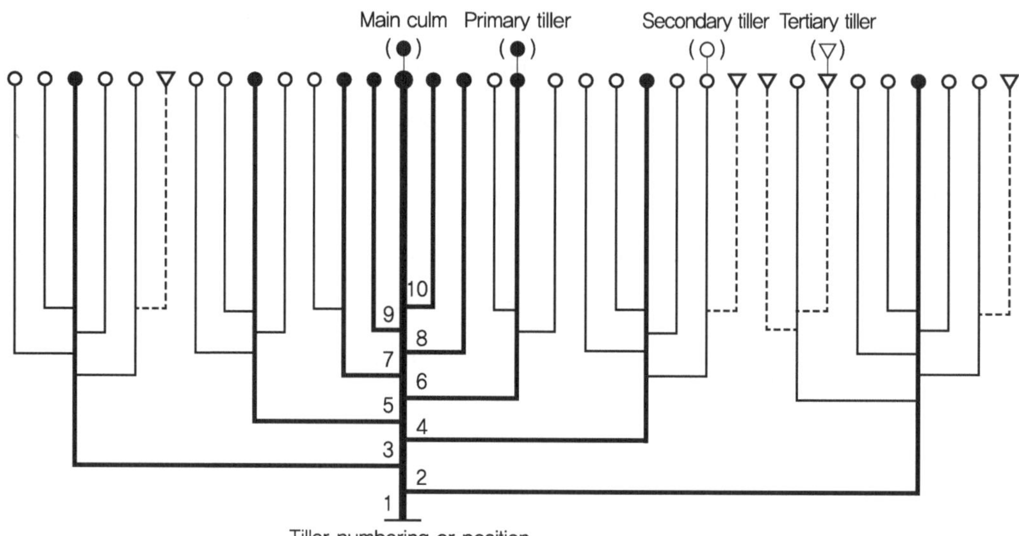

Figure 3-16. Tillering pattern of rice (IRRI, 1981)

B: Diagnostic indicators and methods

① Relative plant height (tiller height) ratio

One week after the maximum tillering stage, tillers on a plant that are at least two-thirds the height of the tallest culm (usually the main culm) are considered productive tillers, whereas those shorter than this threshold are classified as non-productive tillers.

② Leaf emergence rate

According to the theory of synchronous leaf and tiller development, all tillers elongate their leaves in coordination with the main culm. However, after the maximum tillering stage, variation in leaf emergence rate may appear among tillers. If leaf emergence is delayed, the tiller is likely to become non-productive.

③ Timing of tiller emergence

Lower tillers that emerge early tend to develop more biomass and are more likely to become productive tillers, while upper tillers that emerge late have a shorter growth period and limited nutrient availability, often becoming non-productive. In general, tillers that emerge at least 15 days before the maximum tillering stage are considered productive, and those that appear after are considered non-productive.

④ **Number of leaves and root formation**

Tillers with four or more leaves at the maximum tillering stage are typically productive, while those with two or fewer are considered non-productive. According to the synchronous development theory, a tiller with four leaves would have roots emerging from the first leaf node, enabling it to survive independently.

3) Internode Elongation

① At the panicle initiation stage, vegetative growth ceases and reproductive growth begins, initiating the internode elongation process.

② The uppermost internode (peduncle) begins differentiation about 10 days before heading, and its elongation accelerates rapidly 2 days prior to heading, pushing the developing panicle upward through the flag leaf sheath.

③ The peduncle continues elongating for 1~2 days after heading, reaching its maximum length.

④ Lodging (stem bending or breakage) occurs when culm height increases excessively, especially if the basal two internodes elongate, weakening structural support.

⑤ Internode elongation responds strongly to nitrogen fertilization, with excessive nitrogen promoting longer internodes and increasing the risk of lodging.

⑥ During the elongation phase, internode growth can range from 2 to 10cm per day.

4) Panicle Development and Diagnosis

(1) Days to Heading

① The number of days to heading is a key parameter for determining developmental stages (see Table 3~4).

② The timing of panicle development can also be inferred from leaf age index and distance between ligules.

(2) Leaf Emergence and Flag Leaf

① The panicle initiation stage corresponds to the emergence of the fourth leaf before the flag leaf.

Table 3-4. Panicle development stages and estimation of heading date in rice

No.	Development stage	Substage	Leaf number index (%)	Days before heading	Leaf emergence and distance between the arurlicles of the flag leaf and the penultimate leaf	Panicle length (mm)
①	Panicle initiation	-	77	30	Emergence of 4th leaf from the top	0.2
②	Primary branch formation	Primary rachis branch differentiation	81	28	-	0.4
		Secondary rachis branch differentiation	85	26	Emergence of 3rd leaf from the top	1.0
③	Spikelet primordia stage	Spikelet primordium differentiation	87	24	Emergence of 2nd leaf from the top	1.5
		Pistil and stamen differentiation	89	22	-	2.5
		Completion of floret differentiation	92	18	Emergence of top leaf (flag leaf)	10
④	Gameto-genesis	Pollen mother cell formation	95	16	-	30
⑤	Booting stage (Meiosis)	Start of meiosis	96	15	Distance between the arurlicles of the flag leaf and the penultimate leaf (≤ –10cm)	50
		Middle stage of meiosis	97	12	Distance between the arurlicles of the flag leaf and the penultimate leaf (0 cm)	100
		Late stage of meiosis	98	5	Distance between the arurlicles of the flag leaf and the penultimate leaf (≥ +10cm)	150
⑥	Pollen maturation stage	Pollen maturation completed	99	2	Distance between the arurlicles of the flag leaf and the penultimate leaf (≥ +12cm)	Fully elongated
⑦	Heading stage	-	100	0	-	Fully elongated

② At the flag leaf emergence stage, floral primordia begin to differentiate and progress to the formation of pollen mother cells.

③ In cultivars with known leaf number patterns, the total leaf count on the main culm can be used to back-calculate the developmental stage of the panicle based on the current leaf position relative to the flag leaf.

(3) Leaf Number Index (LNI)

① Leaf number index refers to the percentage of emerged leaves relative to the total leaf number of a given cultivar. It is a numerical index that indicates leaf emergence progress.

② Although total leaf number (TLN) can vary depending on cultivar and cultivation conditions, it is typically assumed to be constant (commonly 16 leaves) for standard calculation.

③ Early transplanting tends to increase the TLN by 1~2 leaves, whereas late transplanting generally reduces TLN.

④ LNI values are approximately 77 at panicle initiation, 90 at spikelet differentiation, 97 at meiosis, and close to 100 when pollen maturation begins.

⑤ An LNI of 100 indicates that the flag leaf has fully elongated and that pollen exine formation has commenced.

⑥ For cultivars with TLNs significantly different from the standard 16 (either higher or lower), correction must be applied when calculating the LNI.

(4) Calculation and Correction of LNI

① LNI is calculated as:

(Number of emerged leaves at a given stage / TLN) × 100

② The correction factor is calculated as:

(100 – LNI) × (16 – TLN) / 10

③ Correction is applied in cases where there are too few tillers or excessive non-effective tillers, converting the LNI to a 16-leaf-equivalent value for more accurate growth stage estimation.

④ The corrected LNI is the sum of the raw LNI and the correction factor.

For example, if the TLN is 14 and the number of emerged leaves is 12.6, then:

LNI = 12.6 / 14 × 100 = 90

Correction factor = (100 – 90) × (16 – 14) / 10 = 2

Final corrected LNI = 90 + 2 = 92

(5) Distance between Auricles of Flag Leaf and Penultimate Leaf

① This is a method to estimate panicle developmental stage based on the distance between the auricle of the flag leaf and the auricle of the penultimate leaf.

② The distance between two auricles method is useful for estimating the timing of meiosis, which is highly sensitive to environmental stresses (e.g., temperature).

③ A distance between two auricles of –10cm indicates the start of meiosis, 0cm represents peak meiosis, and +10cm corresponds to the end of meiosis.

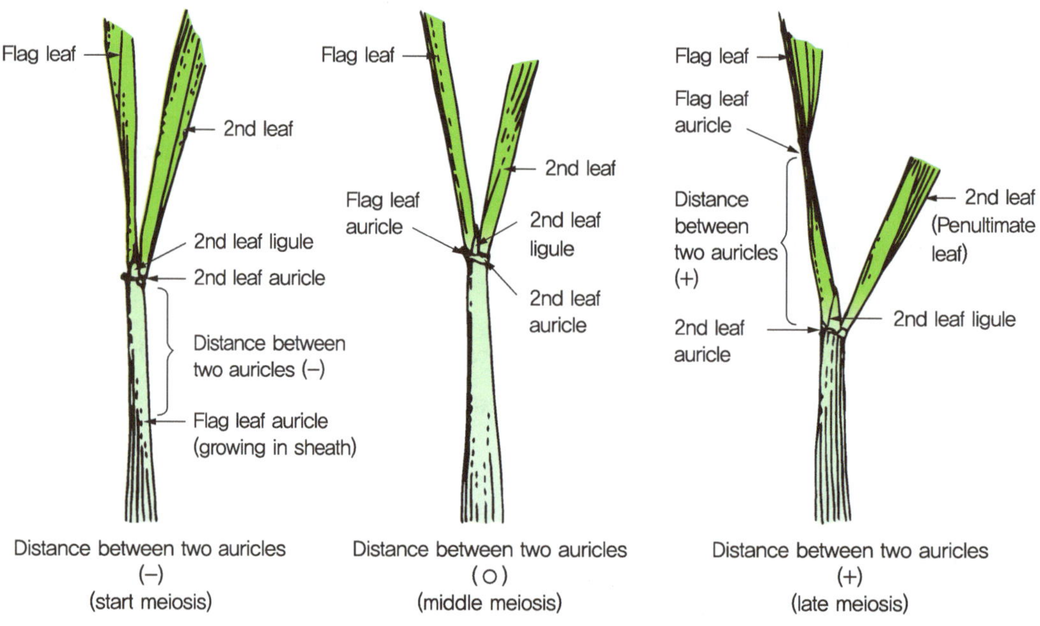

Figure 3-17. Diagnosis of meiosis stage based on leaf distance between two auricles (Matsushima et al., 1970)

7. Panicle Development

1) Transition in Growth Phase

① The transition from the vegetative phase to the reproductive phase in rice is referred to as growth phase transition (also called phase change).

② During the vegetative phase, organs such as leaves, stems, and roots are formed. Once the plant enters the reproductive phase, floral primordia begin to differentiate.

③ This transition is greatly influenced by environmental factors such as photoperiod, temperature, and nutritional status.

④ If the rate of leaf emergence slows down from one leaf every 4~5 days to one every 7~8 days, it is a sign that the plant is transitioning from the vegetative growth phase to the reproductive growth phase.

⑥ The internodes begin to elongate around the time of panicle initiation, so this stage marks the transition from vegetative to reproductive growth.

⑦ Differentiation of the panicle node (neck node) initiates the formation of the panicle, typically occurring at a Leaf Number Index (LNI) of around 76~78.

2) Development of the Panicle

(1) Panicle Differentiation

① The panicle originates from the panicle primordium, which begins to differentiate approximately 30 days prior to heading.

② Following the formation of the flag leaf primordium at the shoot apex, lateral branching primordia develop, giving rise to primary rachis branches, secondary branches, and spikelet primordia.

③ Panicle differentiation typically begins when tiller production ceases. However, under conditions such as heavy nitrogen fertilization in early-maturing cultivars or cultivation in cold regions, tillering may continue even after panicle initiation has begun, resulting in an overlap of vegetative and reproductive growth phases.

(2) Relationship Between Maximum Tiller Stage and Panicle Initiation

The relationship varies by region, cultivation method, and cultivar maturity group, and can be classified into three types:

① In the type where panicle initiation precedes the maximum tillering stage, the panicle begins to differentiate before the plant reaches its maximum tiller number. This pattern is commonly observed in early-maturing cultivars.

② In the type where panicle initiation coincides with the maximum tillering stage, both events occur at the same time, which is typically seen in medium-maturing cultivars.

③ In the type where maximum tillering precedes panicle initiation, the plant reaches its maximum tiller number before panicle initiation begins, a pattern characteristic of late-maturing cultivars.

(3) Occurrence of Sterile Spikelets

① During the booting and heading stages, if the temperature remains below 17°C for several days in temperate *japonica* cultivars or below 20°C in *indica* cultivars, a high incidence of sterile spikelets occurs. This phenomenon is referred to as injury-type chilling damage.

② Such sterility frequently occurs in high-latitude regions of temperate zones and high-altitude regions of tropical zones.

3) Pollen and Embryo Sac Formation

(1) Pollen Formation

① In the pollen primordium, archesporial cells divide to form sporogenous cells, which then develop into microsporocytes (pollen mother cells).

② Microsporocytes undergo meiosis, producing four haploid cells called tetrads, which later develop into pollen grains.

③ Mature pollen grains, yellowish in color, are formed the day before anthesis.

④ Within the anther, pollen mother cells undergo meiosis to form tetrads composed of four pollen cells.

⑤ These cells undergo two mitotic divisions to produce mature pollen, consisting of one pollen tube cell and two sperm cells.

⑥ The pollen tube cell extends into a pollen tube, enabling the sperm cells to reach the embryo sac.

(2) Embryo Sac Formation

① The ovule primordium differentiates into an ovary wall primordium, which develops to enclose the nucellus primordium, eventually forming the ovary.

② A cell located beneath the outer epidermis at the apex of the nucellus enlarges and differentiates into a megaspore mother cell.

③ The megaspore mother cell undergoes three rounds of mitosis, forming the embryo sac.

④ The embryo sac is completed on the day before flowering, and the egg cell attains fertilization competence at this time.

⑤ One megaspore mother cell (2n) undergoes meiosis, forming four haploid megaspores (n).

⑥ Of the four megaspores, three degenerate, and the remaining one undergoes three mitotic divisions to develop into the embryo sac.

⑦ The embryo sac consists of eight nuclei: one egg cell and two synergids at the micropylar end, three antipodal cells at the chalazal end, and two polar nuclei in the center. The synergids and antipodals eventually degenerate.

⑧ The micropyle is located at the lower part of the embryo sac and serves as the entry point for the pollen tube during fertilization.

8. Heading and Flowering

1) Heading

① Heading refers to the emergence of the panicle from the flag leaf sheath. About 1~2 days before heading, internode elongation is nearly complete, and the tip of the panicle reaches the upper portion of the flag leaf sheath. When elongation of the panicle node (neck internode) is finished, the panicle emerges from the sheath.

② Heading begins either during the night before or early in the morning. Once the young panicle is exposed to the external environment, the lemma and palea of the spikelets begin to dry and harden, and the rachis becomes firm and erect.

③ Even after heading and flowering, the neck internode continues to elongate for 1~2 more days, extending the panicle outward by about 10~20cm. Cool temperatures generally reduce the degree of panicle exertion.

④ Within a single rice hill, heading begins with the primary tiller, followed by secondary and then tertiary tillers. It takes about 7 days from the first to the last tiller to complete heading, with over 70% of the panicles emerging during the first 3 days.

⑤ When the number of effective tillers per hill is high (e.g., around 30), the heading period may extend up to 10 days, and with fewer tillers, heading may complete in 4~5 days. Thus, heading duration varies depending on the number of tillers per hill.

2) Flowering

(1) Characteristics of Flowering

① Flowering occurs when the tips of the lemma and palea of a spikelet slightly open, allowing the anthers to extrude. As rice is a self-pollinating crop, pollination is usually completed just before anthesis.

② In rice, spikelets at the apical part of the panicle begin to flower soon after emergence, so heading and flowering occur simultaneously.

③ Within a panicle, flowering begins at the upper primary branches and progresses downward to lower branches, following the developmental sequence of spikelets.

④ Within a single branch, flowering typically begins at the terminal spikelet, progresses downward, then slightly reverses upward. The second spikelet from the tip is usually the last to flower.

⑤ The flowering order within a panicle matches the order of spikelet differentiation, and also corresponds to the order of grain filling and ripening.

⑥ Flowering starts early in the morning (around 7:00 a.m.), with the peak flowering period occurring between 8:00 and 11:00 a.m., during which most spikelets flower.

(2) Environmental Conditions Affecting Flowering

A: Temperature

① The minimum temperature for flowering is 15°C, the optimal range is 30~35°C, and the upper threshold is 50°C. Temperatures above 35°C can result in increased spikelet sterility.

② Under high-temperature conditions, flowering is concentrated between 9:00 and 11:00 a.m., whereas under low-temperature conditions, flowering occurs later, typically between 12:00 and 5:00 p.m.

B: Light

① Under optimal temperatures (~30°C), artificial light exposure from the night before can induce continuous flowering starting around midnight, even in the absence of natural light.

② If plants are kept in darkness during the day and exposed to light at night, the flowering time shifts accordingly and occurs at night.

C: Rain and wind

① When it rains during flowering, lodicules fail to swell, preventing spikelet opening. However, most spikelets will open once the rain stops.

② As the filaments elongate inside the floret and push the anthers against the inner wall of the lemma and palea, the pressure causes them to burst, releasing pollen. This leads to cleistogamous pollination (pollination without full floret opening).

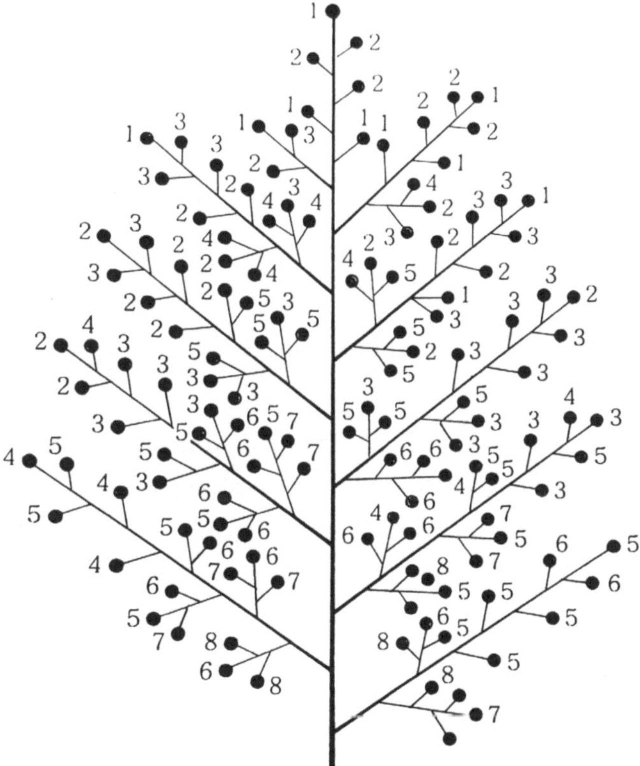

Figure 3-18.
Flowering sequence of a rice panicle
(The numbers indicate the dates on which individual spikelets flowered.)

3) Fertilization and Pollination

(1) Pollination

① During rice anthesis, just before the lemma and palea open, the anthers burst and release large amounts of pollen onto the stigma of the same flower, resulting in self-pollination. Rice is predominantly self-pollinating, with a cross-pollination rate of less than 1%.

② About 1~2 hours before or after spikelet opening, the lodicules begin to desiccate, causing the palea to close. Withered anthers remain enclosed within the closed spikelet and eventually disintegrate. This process is called cleistogamy.

③ The stigma is structurally well-adapted to receive a large number of pollen grains but withers shortly after the flowering period ends.

④ Within 2~3 minutes after pollen deposition, the pollen tube emerges through the germ pore and begins to elongate into the stigma tissue. Simultaneously, the pollen tube cell forms the tube structure. Pollen grains lose their viability about 5 minutes after adhering to the stigma.

⑤ Each pollen grain contains two sperm nuclei and one vegetative nucleus (also called the pollen tube nucleus). As the pollen tube elongates, the cytoplasmic contents of the pollen grain are transported into the tube.

⑥ The number of pollen grains that are fertilized on the stigma is in the hundreds to thousands, but the pollen grains that are finally fertilized and form seeds germinate the fastest, and only the one that arrives at the stigma first enters the sac. However, if there is little pollen grain attached to the stigma, the germination or elongation of the pollen tube tends to be delayed.

⑦ The optimal temperature for pollen germination is 30~35°C. Abnormalities in pollen germination occur at low temperatures below 20°C and high temperatures above 40°C. Excessive humidity or dryness inhibits germination. Pollen germination and fertilization occur regardless of light conditions.

(2) Fertilization

① The pollen tube elongates downward along the style, enters the ovule through the micropyle by following the ovary wall and nucellus, and finally reaches the embryo sac.

② The time required for the pollen tube to reach the embryo sac is about 15 minutes in optimal conditions, 30 minutes on average, and up to 60 minutes under low temperatures.

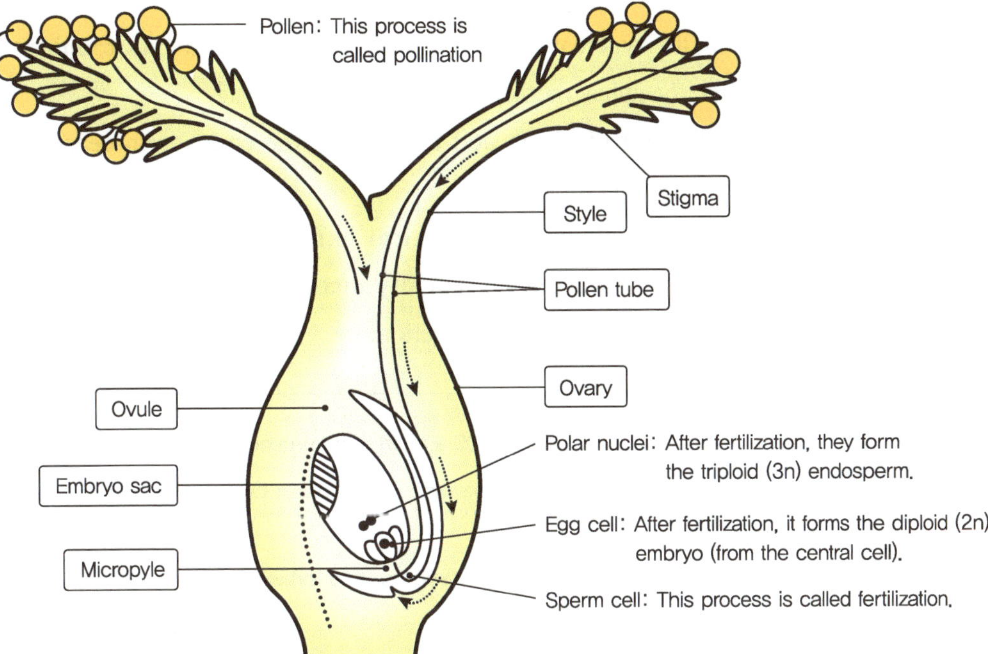

Figure 3-19. Pollination and fertilization in rice (Yusa Goto et al., 2013)

③ Upon arrival, the tip of the pollen tube penetrates a synergid cell and ruptures between the egg cell and the polar nuclei, releasing two sperm nuclei, enabling fertilization.

④ One sperm nucleus (n) fuses with the egg cell (n) to form the diploid embryo (2n), while the other sperm nucleus (n) fuses with the polar nuclei (2n) to form the triploid endosperm (3n).

⑤ This process, where two fertilization events occur within the ovary, is called double fertilization. One fertilization produces the embryo from the sperm and egg cells, and the other forms the endosperm from the fusion of the sperm nucleus with the two polar nuclei.

⑥ The total time required from pollination to the completion of fertilization is approximately 5 to 6 hours, assuming an average daily temperature of 28°C.

⑦ The optimal temperature for fertilization is 30~32°C. At temperatures above 35°C, fertilization failure may occur. When the temperature is below optimal, the fertilization process is prolonged, and at temperatures below 20°C, fertilization does not proceed.

(3) Morphogenesis of the Embryo and Endosperm

① After double fertilization, the zygote undergoes its first division the next day, forming two cells. Around 4 days later, the apical meristem begins to differentiate.

② Around 9 to 10 days after fertilization, cell division in the endosperm ceases, and endosperm cell determination is completed.

③ By 11 to 12 days after fertilization, morphological differentiation of the embryo is complete. Thereafter, the embryo physiologically matures and is considered fully developed by approximately 25 days after fertilization.

① The flowering of the floret typically begins around 8:00 a.m. in most rice cultivars.

② As filaments elongate, anthers touch the lemma, release pollen upon bursting, and the spikelet then closes. It does cleistogamy.

③ Numerous pollen grains are deposited on the feathery stigma, and fertilization occurs as the pollen tubes grow through the style into the embryo sac.

④ During anthesis, pollen grains can be observed streaming down.

⑤ The lemma and palea open, allowing effective pollination to take place.

Figure 3-20. Floral opening (anthesis) and fertilization in rice

4) Accumulation of Storage Compounds

Storage compounds transported to the brown rice enter through the vascular bundles of the rachilla (small branch of the panicle) and are delivered via the dorsal vascular tissues of the ovary wall. Since the endosperm tissue lacks vascular structures, imported compounds must pass cell-to-cell and are first delivered to the innermost cells of the endosperm, where they are transformed into storage-appropriate forms and then accumulated.

(1) Starch

① The starch content in the stem and leaf sheath remains high until heading, but decreases during the grain-filling stage.

② Most of the compounds delivered to the endosperm are water-soluble carbohydrates, primarily sucrose. More than 90% of storage compounds in the endosperm are stored in the form of starch granules.

③ Around 4~5 days after flowering, numerous fine granules begin to form in the cytoplasm of central endosperm cells, and starch granules (amyloplasts) begin to accumulate starch within them.

④ Around 5~6 days after fertilization, the central region of the endosperm begins to develop. Starch accumulation starts from the innermost cells of the endosperm and progresses outward. By around 15 days after fertilization, starch accumulation is completed in the inner endosperm, and by 30~35 days, it reaches the subaleurone region.

⑤ Amyloplasts are formed in storage tissues. In non-glutinous rice, amylose content is around 23%, while in glutinous rice, amylopectin predominates and amylose is about 9%.

⑥ Around 15 days after fertilization, large starch granules with diameters of about 30~40 μm can be observed in the central endosperm.

⑦ A rice grain consists of approximately 5% embryo (germ), 65% endosperm (white rice), and 30% bran layers, including aleurone layer, seed coat, pericarp, and rice bran.

(2) Aleurone Layer

① The outermost layer of the endosperm completes cell division about 10 days after fertilization, and differentiates into the aleurone layer.

② Aleurone cells accumulate protein bodies and lipid bodies, collectively known as aleurone grains.

(3) Protein

① The storage protein content in the endosperm is about 6~8%, stored as protein bodies in the cytoplasm.

② Around 6~7 days after flowering, protein bodies become visible, reaching a maximum around 20 days. These granules are 1~3 μm in diameter.

③ Protein accumulation is generally higher in upland rice than in paddy rice, and even for the same cultivar, upland cultivation increases protein content.

④ Top-dressing with nitrogen fertilizer after heading increases the grain protein content.

⑤ The major storage protein is glutelin (70~80%), followed by prolamin.

⑥ Smaller grains with lower 1,000-grain weight typically have higher protein concentrations.

(4) Lipid

The lipid content in rice grain is about 2~3%, with approximately 30% located in the embryo, and most of the remainder stored in lipid bodies of aleurone layer cells in the endosperm.

5) Grain Maturation (Ripening)

Grain maturation involves the formation of the embryo and endosperm cells, and the translocation of starch reserves. Approximately 20~30% of starch accumulated in the stem and leaf sheath before heading and 70~80% of photosynthates produced after heading are mobilized to the panicle for grain filling. During early grain development, endosperm cell numbers increase, and in the later stage, storage compounds accumulate in differentiated cells.

(1) Caryopsis Development

A: Grain development process

① Grain (caryopsis) development progresses first in length, then in width, and finally in thickness.

② Grain length is completed around 5~6 days after fertilization, and width reaches its full extent by 15~16 days.

③ Grain thickness is completed by 20~25 days, and by 25 days, the overall shape of brown rice (unpolished grain) is established. At this stage, the pericarp still contains chlorophyll, giving the grain a green appearance and enabling photosynthesis. Internally, grain filling continues.

④ Around 30 days after fertilization, the grain slightly shrinks in size, and the chlorophyll in the pericarp disappears. By 45 days, the grain reaches full maturity.

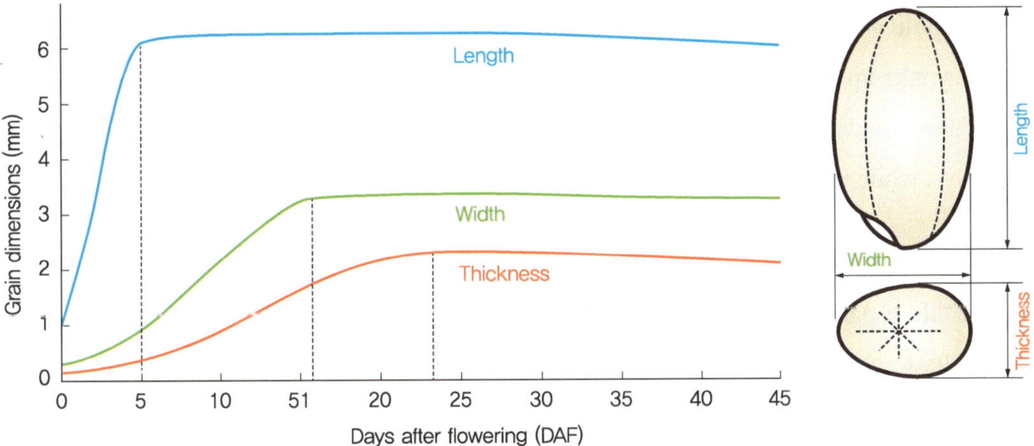

Figure 3-21. External morphological development of brown rice (Hoshikawa, 1975)

B: Developmental patterns of brown rice (Unpolished grain)

① The fresh weight of brown rice increases almost linearly until about 20 days after fertilization, reaching its maximum around 25 days, and slightly decreases after 35 days.

② Dry matter accumulation increases markedly between 10 to 20 days after heading and is nearly complete by around 35 days.

③ Moisture content is initially high during the early grain-filling stage but continuously declines until the full maturity stage, reaching approximately 20% by around 35 days after heading.

(2) Maturation of the Panicle

A: Panicle morphology

① At the time of heading and flowering, both the rachis (main panicle axis) and the secondary rachis branches (rachilla) remain upright.

② From about 4 to 5 days after fertilization, starch begins to accumulate, initiating the grain-filling process.

③ Around 30 days after heading, the panicle tip bends downward, drooping below the flag leaf sheath.

B: Grain-filling duration

① In early sowing of early-maturing cultivars during high-temperature periods, harvesting is possible within 30~35 days after heading. However, under low-temperature conditions, it may take up to 60 days.

② Under standard cultivation conditions, early-maturing cultivars generally require about 40 days, while medium-to-late-maturing cultivars require around 50 days after heading.

C: Grain maturation sequence

① Within a panicle, spikelets on the upper primary branches that flower first also mature earliest, while those on the lower branches that flower later mature more slowly.

② Grains on upper branches tend to be more well-developed and fully filled, while those on lower branches have a higher proportion of poorly filled or inferior grains.

③ Among the spikelets on a branch, the apical spikelet that flowers first matures fastest, followed by the rest in the flowering sequence from base to tip.

④ The second spikelet from the tip tends to develop the slowest, resulting in lighter grain weight and higher likelihood of being incomplete or sterile, especially under poor weather conditions during the grain-filling stage.

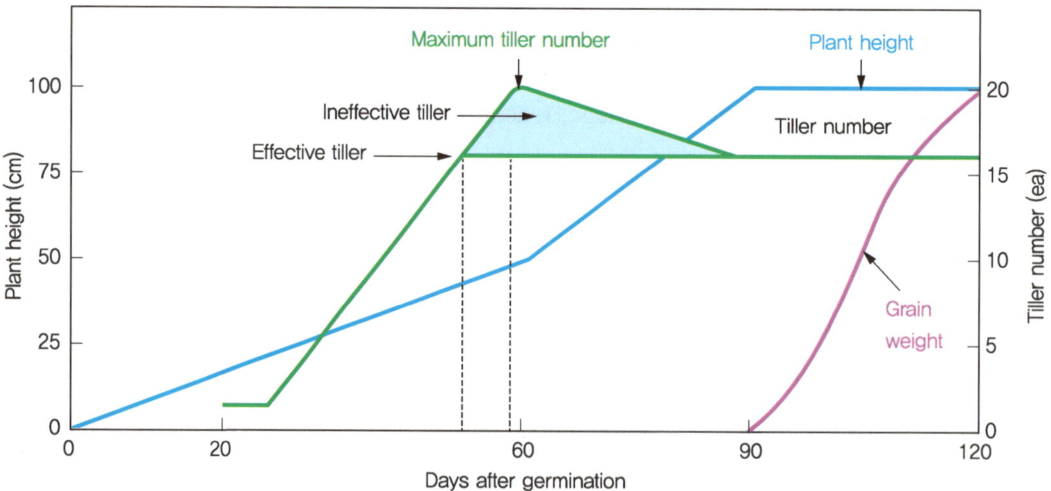

Figure 3-22. Changes in biomass accumulation during rice growth

(3) Grain Filling and Environmental Conditions

A: Temperature

① The optimum temperature for ripening is 20~22°C, and among meteorological factors, air temperature most strongly affects the translocation of assimilates to the panicle.

② High temperatures around 30°C during the grain-filling period generally shorten the maturation duration and degrade grain quality.

③ In the early ripening stage, the embryo and endosperm develop actively, and photosynthesis in the leaves is vigorous, so high solar radiation and moderately high temperatures are advantageous.

④ In the late ripening stage, slightly cooler temperatures are more favorable for the translocation and accumulation of storage materials than high temperatures.

⑤ The proliferation of endosperm cells occurs more rapidly at temperatures near 30°C, and is delayed at temperatures below 20°C.

⑥ The amount of carbohydrate translocation to the panicle increases with temperature within the range of 17~29°C. However, below 17°C, translocation to the panicle is significantly suppressed.

⑦ A relatively large diurnal temperature range is favorable for grain filling. If nighttime temperatures are high, a large portion of carbohydrates produced by photosynthesis is consumed by respiration, reducing the amount translocated to the panicle.

⑧ The optimal temperature regime for grain filling in rice is approximately 29°C during the day and 19°C at night for the first 10 days after heading, and 25°C (day) and 15°C (night) thereafter.

⑨ In tropical regions, rice yields tend to be lower than in temperate zones primarily due to excessively high temperatures during the ripening stage and smaller diurnal temperature variation.

⑩ When temperatures during grain filling exceed 23°C, the early ripening stage may produce a higher 1000-grain weight compared to optimal temperatures (20~22°C), but the late ripening stage results in lower grain weight. This is because high temperatures accelerate senescence and interfere with the translocation of assimilates.

Table 3-5. Minimum, optimum, and maximum temperatures for each growth stage of rice (Source: Yoshida, 1977)

No.	Growth stage	Minimum temperature (°C)	Optimum temperature (°C)	Maximum temperature (°C)
①	Germination stage	10	20~35	45
②	Seedling emergence stage	12~13	25~30	35
③	Rooting/Establishment stage	16	25~28	35
④	Leaf development stage	7~12	31	45
⑤	Tillering stage	9~16	25~31	33
⑥	Panicle initiation preparation	15	-	-
⑦	Panicle initiation stage	15~20	-	38
⑧	Flowering stage	22	30~33	35
⑨	Ripening (maturation) stage	12~18	20~25	30

B: Solar radiation and sunshine duration

① Lack of solar radiation reduces photosynthetic activity, which negatively impacts grain filling in rice.

② Approximately 20~30% of carbohydrates accumulated in the panicle are derived from assimilates stored in the culm and leaf sheath before heading, while the remaining 70–80% are synthesized through photosynthesis after heading.

③ High levels of solar radiation and long sunshine duration during grain filling are directly correlated with increased yield.

C: Nutrients

① Nitrogen (N) plays a critical role in determining photosynthetic rate during grain filling. Leaves with high nitrogen content exhibit higher photosynthetic capacity.

② Excess nitrogen application at the panicle initiation stage can lead to excessive vegetative growth, increasing the leaf area index (LAI), which worsens the light interception posture of the population and reduces photosynthetic efficiency. Therefore, reduced nitrogen management is needed at this stage.

③ Nitrogen has a significant impact on grain filling. Topdressed N fertilizer (spikelet fertilizer) reduces the formation of sterile spikelets and enhances the photosynthetic performance of leaves before heading, promoting grain development. In particular, grain nitrogen fertilizer (post-heading application) increases chlorophyll content in the functional leaves after heading, thereby enhancing photosynthesis and improving grain filling.

④ Nitrogen deficiency decreases photosynthetic activity and lowers the 1000-grain weight, while excess nitrogen reduces the sugar concentration in leaves during early ripening, limiting assimilate translocation to the panicle and thus reducing grain filling.

⑤ The threshold nitrogen concentration in the flag leaf required to maintain high grain weight is approximately 1.2%, and about 0.9% at harvest.

⑥ Phosphorus (P) is rapidly translocated to the panicle during grain development and is accumulated as phytic acid in the aleurone layer granules.

D: Typhoons

① Water stress during grain filling results in a higher proportion of immature kernels, but light typhoons may provide beneficial rainfall to alleviate moisture deficiency.

② If strong winds occur 3~5 days after heading, the panicle may desiccate, leading to white immature kernels (chalky grains), and the risk of diseases such as rice blast or panicle blast increases, ultimately reducing grain yield.

③ In Korea's monsoonal climate, 2~3 typhoons commonly impact the southern regions during the grain filling season, causing varying degrees of damage.

9. Photosynthesis and Respiration

1) Photosynthesis

(1) Characteristics of Photosynthesis

A: Photosynthesis

Photosynthesis is a physiological process in which green plants utilize light energy to synthesize carbohydrates (CH_2O) from water (H_2O) and carbon dioxide (CO_2):

$6H_2O + 6CO_2 \rightarrow C_6H_{12}O_6 + 6O_2$.

The sugars produced through photosynthesis are the primary precursors for synthesizing carbohydrates, lipids, and proteins, thereby providing the organic energy essential for plant metabolism.

Table 3-6. Photosynthetic characteristics of rice (*Oryza sativa* L.)

No.	Photosynthetic factor	Measured value	Experimental conditions
①	Light compensation point	0.4~1 klux	25°C, 300ppm CO_2
②	Light saturation point	45~60 klux	30°C, 300ppm CO_2
③	Optimum temperature	20~33°C (*Japonica*) / 25~35°C (*Indica*)	300ppm CO_2, 50 klux
④	CO_2 compensation point	55ppm	25°C, ≥10 klux
⑤	Net photosynthetic rate	40~50mg CO_2/dm²/h	Optimum temperature, light saturation conditions
⑥	Biochemical pathway	C_3 pathway	-
⑦	Bundle sheath cells	Absent	-
⑧	Photorespiration	Present	-

*The atmospheric CO_2 concentration in 2022 was 423 ppm.

B: Photosynthesis

Photosynthesis primarily takes place in the leaf blade (lamina), but it also occurs in other green tissues such as the leaf sheath, culm (stem), and panicle, where chloroplasts containing chlorophyll are present.

C: Carbon fixation

Carbon fixation refers to the process in which carbon from atmospheric CO_2 is incorporated into sugars. During photosynthesis, water molecules (H_2O) are split (photolysis), resulting in the oxidation of water as it loses electrons (e^-) and hydrogen ions (H^+). In turn, carbon dioxide (CO_2) is reduced by accepting these hydrogen ions and electrons to form glucose ($C_6H_{12}O_6$). As a by-product of this process, oxygen (O_2) is released into the atmosphere.

D: Nourishment

The carbohydrates produced via photosynthesis can be used by the plant as an energy source through cellular respiration, or as building blocks for synthesizing structural compounds such as cellulose. Most crops produce more sugars than immediately needed, and the excess is converted into starch and stored in seeds, storage roots, tubers, or fruits. The organic matter generated by the plant not only supports its own growth and development but also serves as a primary energy and nutrient source for other organisms in the ecosystem.

Table 3-7. Comparative photosynthetic characteristics of C_3 and C_4 crops

No.	Parameter	C_3 Plants	C_4 Plants	Reference
①	Average water requirement (transpiration coefficient)	671 (Rice: 300) mm H_2O/kg dry matter	318mm H_2O/kg dry matter	Downes (1969)
②	Maximum growth rate	34~39g/m^2/day	50~54g/m^2/day	Monteith (1978)
③	Optimum light intensity	30~50% of full sunlight	Full sunlight	Hatch (1973)
④	Optimum temperature for photosynthesis	15~30°C	30~45°C	Hatch (1973)
⑤	Photosynthetic rate per unit leaf area	50%	100%	Hatch (1973)
⑥	Water use efficiency (WUE)	1.49mg/g	3.14mg/g	Downes (1969)

E: Chloroplasts

Chloroplasts are abundant in the mesophyll tissue of leaves and serve as the cellular organelles where photosynthesis takes place. Their main structural and functional characteristics are as follows:

① The chloroplast, enclosed by a double membrane consisting of an inner and outer membrane, is composed of stroma, a fluid matrix containing photosynthetic enzymes, and thylakoids, which are membrane-bound structures containing chlorophyll and components of the electron transport chain. A stack of 10 to 20 thylakoids forms a structure known as a granum (plural: grana).

② During the light reactions of photosynthesis, light energy absorbed by chlorophyll in the thylakoid membranes is converted into chemical energy in the form of ATP and NADPH.

③ The dark reactions, or the Calvin cycle, occur in the stroma where ATP and NADPH are used to fix carbon dioxide (CO_2) into sugars.

F: Mitochondria

Mitochondria are cellular organelles responsible for converting the chemical energy of organic compounds into ATP through cellular respiration.

① Mitochondria are composed of folded inner membranes known as cristae, which house the electron transport chain and ATP synthase, and a fluid matrix that contains respiratory enzymes. The cristae increase surface area for efficient ATP production.

② Cellular respiration begins with glycolysis, which occurs in the cytoplasm and breaks down glucose into two molecules of pyruvate. Subsequent stages, including the Krebs cycle (also called the citric acid cycle) and the electron transport chain, take place within the mitochondria.

③ In the Krebs cycle, pyruvate is completely oxidized to carbon dioxide (CO_2), and hydrogen carriers such as NADH and $FADH_2$ are produced.

④ The electron transport chain accepts electrons from NADH and $FADH_2$ and transfers them through a series of carriers, ultimately generating ATP.

G: Temperature coefficient (Q_{10}) in photosynthesis

The temperature coefficient Q_{10} in photosynthesis indicates that for every 10 °C increase in temperature, the physiological activity approximately doubles. Q_{10} is calculated using the formula:

Q_{10} = (Rate at T+10 °C) / (Rate at T °C)

(2) Internal Factors Affecting Photosynthesis

① The photosynthetic capacity per unit leaf area in rice varies even under identical external conditions due to cultivar differences, leaf aging, water content, chlorophyll concentration, mineral nutrient levels, and the quantity and activity of photosynthetic enzymes.

② In a rice canopy, even if individual leaf photosynthesis rates are equal, differences in leaf position, angle, and planting density can affect light interception, temperature, and CO_2 concentration, ultimately altering total canopy photosynthesis.

③ An efficient translocation of assimilates (photosynthates) enhances the photosynthetic rate.

④ To maintain optimal photosynthetic capacity, rice leaves require minimum mineral nutrient concentrations of nitrogen (N) 2.0%, phosphorus (P_2O_5) 0.5%, magnesium (Mg) 0.3%, and calcium (Ca) 2.0%. Photosynthesis declines when nutrient concentrations fall below these thresholds.

⑤ Increasing nitrogen fertilizer application generally enhances leaf area expansion and improves photosynthetic capacity per unit area, leading to higher photosynthetic output and biomass accumulation per plant.

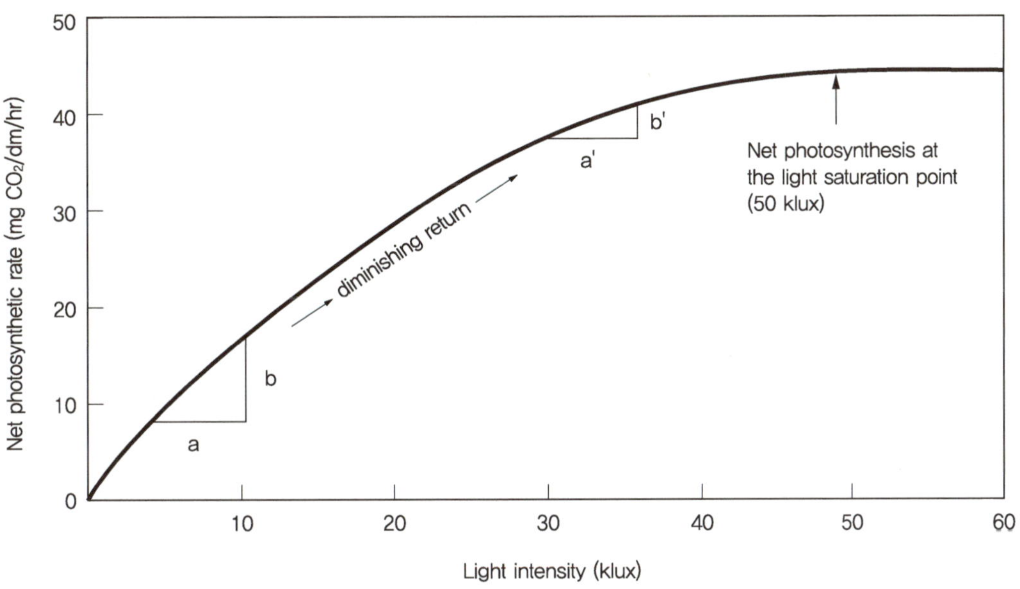

Figure 3-23. Relationship between photosynthesis and respiration in rice during the flowering stage (Source: Cock and Yoshida, 1973)

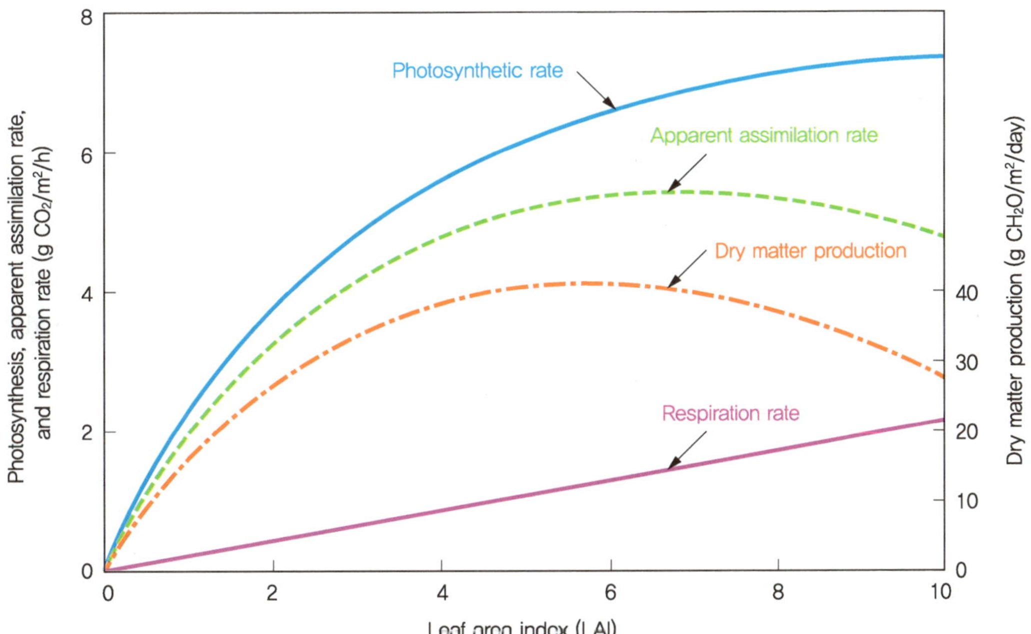

Figure 3-24. Relationship between leaf area index and photosynthesis, respiration, and dry matter production (Source: Tanaka et al., 1966)

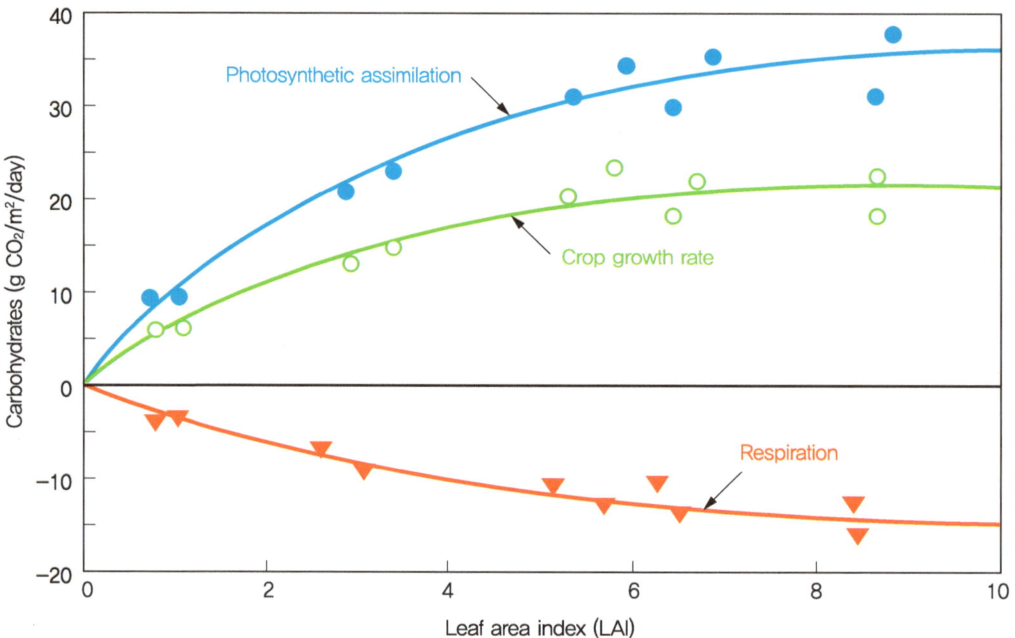

Figure 3-25. Relationship among photosynthesis, crop growth rate, and respiration in rice during the booting stage (Source: Cock and Yoshida, 1973)

(3) External Factors Affecting Photosynthesis

A: Light intensity (at the individual plant level)

① The photosynthetic capacity per unit leaf area in rice reaches the light saturation point at approximately 50 klux. In isolated plants, the photosynthetic rate increases with rising temperature up to the optimum for growth, while the light saturation point tends to decrease.

② During the tillering stage, when light intensity exceeds 30~40% of full sunlight (approximately 30~40 klux out of 120 klux), the photosynthetic rate remains relatively constant regardless of further increases in light intensity, indicating that the light saturation point has been reached.

③ As temperature increases up to the optimum level for growth, the photosynthetic rate increases, and the light saturation point decreases.

④ When the atmospheric CO_2 concentration increases, the photosynthetic rate also increases, and the light saturation point shifts higher.

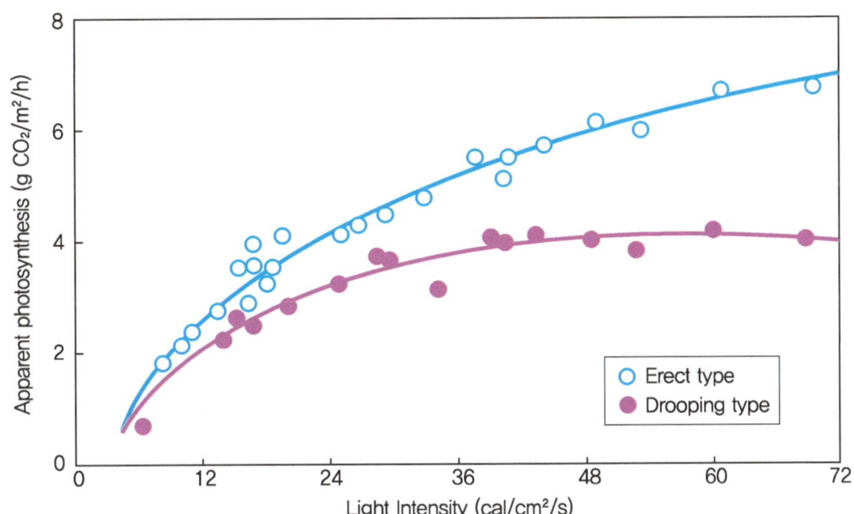

Figure 3-26. Apparent photosynthesis at varying light intensities in rice with erect and drooping leaf types (Source: Tanaka et al., 1969)

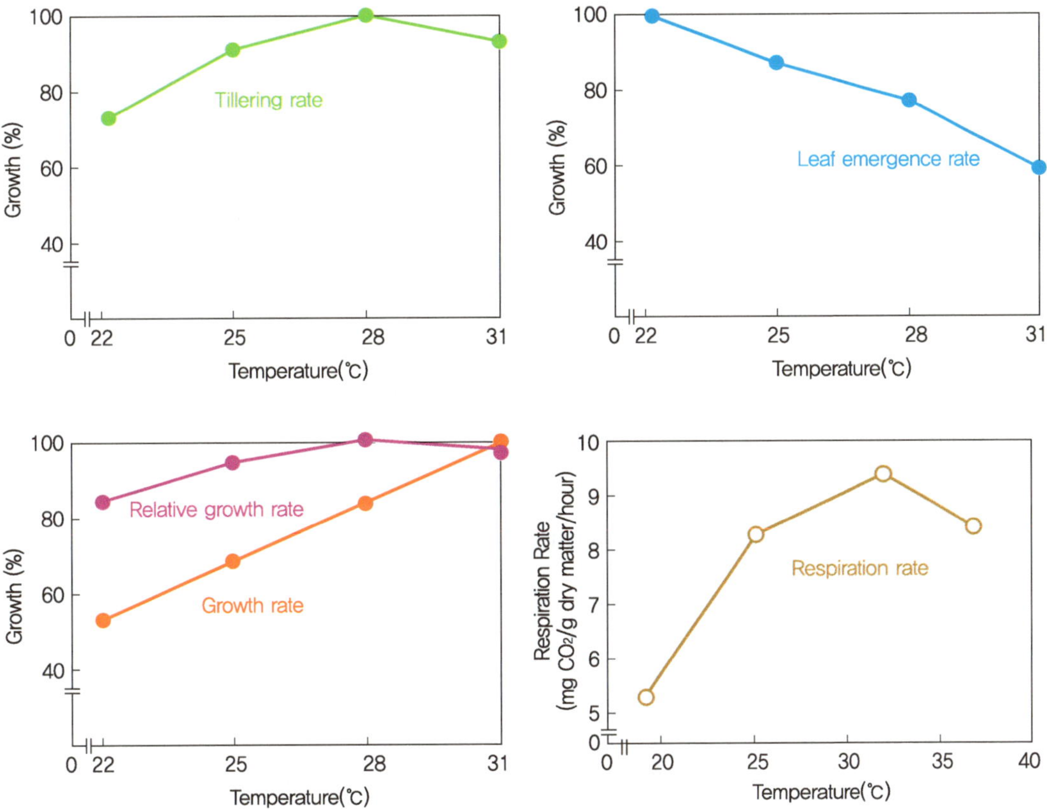

Figure 3-27. Changes in growth parameters of rice under different air temperatures in laboratory conditions (Source: Yoshida, 1973)

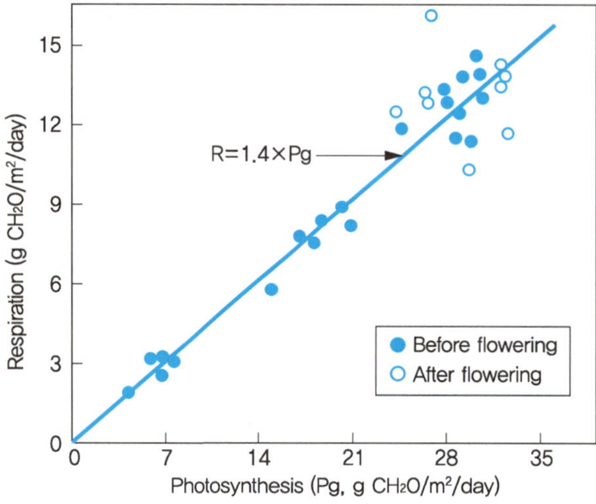

Figure 3-28.
Relationship between photosynthesis (Pg) and respiration (R) in rice before and after flowering (Source: Cock and Yoshida, 1973)

B: Temperature

① In rice, apparent photosynthesis does not vary significantly within the temperature range of 18~34°C. This is because true photosynthesis increases with temperature, but so does respiration. In other words, within the cultivable temperature range, higher temperatures result in increased respiration.

② Below 18°C, photosynthetic activity in rice declines markedly. Although the optimum temperature range for photosynthesis is approximately 20~33°C, higher temperatures also raise respiration. Therefore, dry matter production (i.e., apparent photosynthesis) tends to be higher at relatively cooler temperatures around 20~21°C.

③ Higher temperatures are advantageous under low light, whereas under extremely high temperatures (≥35°C), lower light intensity is more favorable.

④ The light saturation point of rice leaves is higher in cooler climates, meaning stronger solar radiation is required than in warmer regions.

C: Light intensity (Canopy conditions)

① During the maximum tillering stage, at least 60% of full sunlight (approximately 70 klux) is required for optimal photosynthesis.

② In dense canopies where lower leaves receive insufficient light, photosynthesis continues to increase up to 80 klux. (In Korea, the maximum solar radiation during midsummer can reach about 120 klux.)

③ As canopy density increases, the light intensity required to reach the saturation point also increases. However, with greater leaf area, respiration increases proportionally, which may result in a decrease in net assimilation.

④ The canopy-level light saturation point varies with planting density, growth stage, and leaf area index (LAI). In actual field conditions, the canopy often fails to reach the saturation point during stages when LAI is high.

D: CO_2 concentration

① At an atmospheric CO_2 concentration of 400 ppm, rice achieves only about 50% of its maximum photosynthetic potential. As CO_2 concentration increases, photosynthesis also increases, but this response plateaus at around 2,000 ppm, known as the CO_2 saturation point.

② Moderate wind, such as a gentle breeze, facilitates CO_2 supply to leaves and thereby enhances photosynthesis.

E: Growth stage

① In rice, photosynthesis typically peaks around 9 a.m. and remains high until approximately 4 p.m., after which it declines sharply.

② Photosynthetic capacity per unit leaf area is highest during the tillering stage and decreases thereafter.

③ Total canopy photosynthesis reaches its peak between the maximum tillering and booting stages when leaf area is greatest.

④ After heading, canopy photosynthesis declines due to senescence of lower leaves, reduction in total leaf area, and leaf aging.

⑤ Shortly after transplanting, when the plant is still in an individual state (not yet forming a canopy), photosynthetic capacity is at its highest.

⑥ During the panicle initiation stage under canopy conditions, total photosynthesis is greatest, and the leaf area index reaches its maximum just before heading.

⑦ The respiration rate of non-leaf organs (such as leaf sheaths and culms) reaches its peak during the heading stage.

(4) Canopy Photosynthesis

A: Increasing rice yield

Enhancing dry matter production through improved photosynthesis is essential for increasing rice yield.

Canopy assimilation capacity can be expressed as:

Canopy Assimilation = Leaf Area (A) × Light Interception Efficiency (F) × Photosynthetic Capacity per Unit Leaf Area (P)

To maximize dry matter production, the optimal combination of these three components (A × F × P) must be achieved.

① Maximize leaf area (A) to absorb as much solar radiation as possible.

② Improve light interception efficiency (F) by ensuring leaf architecture favors effective light capture.

② Enhance photosynthetic capacity per unit leaf area (P) to maximize conversion of absorbed light into photosynthates.

B: Dry matter production

① As leaf area increases, the photosynthetic rate initially increases linearly, but the rate of increase diminishes and eventually plateaus.

② Respiration increases linearly with rising temperature and leaf area.

③ Dry matter production is calculated as the difference between photosynthesis and respiration. When the diurnal temperature range is wide, nighttime respiration decreases, resulting in increased net dry matter production.

C: Light interception and transmission

① Light interception in a rice canopy is influenced by leaf arrangement, leaf angle, and the length and width of the leaf blades.

② In individual plants, leaves must be well-spread to maximize light absorption.

③ In dense canopies, excessively large and drooping upper leaves shade the lower leaves, reducing their ability to intercept light and negatively affecting total canopy photosynthesis.

D: Ideal plant type favorable for photosynthesis

① Under high light conditions, erect-leaf types exhibit greater photosynthetic performance than drooping types. Erect leaves allow deeper light penetration into the canopy, improving light distribution and enhancing overall canopy photosynthesis.

② The canopy light interception posture is improved in erect-leaf cultivars with large leaf inclination angles, increasing light-use efficiency and contributing to higher yield potential.

③ When upper leaves are small, thin, erect, evenly distributed, and non-overlapping, light can reach lower leaves more effectively, resulting in superior whole-canopy light capture and photosynthetic efficiency.

④ Rather than promoting excessive development of upper leaves through over-application of nitrogen fertilizer, applying silicon (Si) fertilizer helps keep leaf blades erect, which improves light interception efficiency even with a large leaf area, ultimately leading to increased yield.

⑤ Tongil-type rice cultivars generally exhibit better canopy light interception posture than *Japonica*-type cultivars.

Table 3-8. Advantages and disadvantages of morphological traits in rice cultivation

No.	Plant part	Trait characteristics	Advantages	Disadvantages
①	Leaf	Thick	It favors an erect posture and has a higher photosynthetic rate per unit area	-
		Thin	Enables higher leaf area index (LAI)	-
		Short and small	Uniform distribution and more erect posture in canopy	-
		Erect	High light-use efficiency including diffused light under high LAI	-
		Drooping	-	Low light-use efficiency
②	Culm	Short and strong	Resistant to lodging	-
		Tall	Suppresses weed growth and is favorable for early growth	Susceptible to lodging
③	Tiller	Erect and strong	Enhances light penetration into the canopy	-
		Few	Favorable for direct seeding	Low initial biomass

③	Tiller	Many	Allows wider planting distance	-
			Compensates for missing hills	-
			Accelerates LAI development (in transplanted rice)	Excessive initial biomass
④	Root	Shallow, high T/R ratio	High aboveground assimilation capacity	-
		Deep, low T/R ratio	Better utilization of soil nitrogen	-
⑤	Panicle	Low sterility under high N	Allows greater nitrogen use	-
		High harvest index	Potential for high yield	-
		Small grains	Faster ripening	-
		Large grains	Potential for high yield	-
⑥	Growth duration	Early-maturing	High productivity relative to growing period	-
			High water use efficiency	-
			Requires dense planting for high yield	-
		Late-maturing	-	May suffer in low-fertility soils
⑦	Photoperiod Sensitivity	High	High regional adaptability	-
		Low	Allows year-round cultivation and multiple cropping systems	-
⑧	Growth rate	Fast	Suppresses weeds	-
			Favorable in early-maturing varieties	-
		Slow	Prevents excessive vegetative growth in late stages	-

(5) Source and Sink

A: Source (Supply organs) and sink (Assimilate-receiving organs)

① The photosynthetic capacity of source leaves increases when the demand for assimilates is high (i.e., sink strength is strong). In contrast, if the sink demand is low, photosynthesis tends to decrease.

② Traits associated with the source include plant architecture, leaf area, chlorophyll content, leaf thickness, photosynthetic capacity, maintenance of leaf function, and root vitality.

③ Sink organs include panicles, spikelets, leaves, and roots. The photosynthetic activity of the plant is influenced by the size and vigor of these sink organs.

B: Source-sink interaction

① No matter how high the photosynthetic capacity of the source leaves, if the sink (e.g., panicles) is small, assimilates are not utilized efficiently, and overall yield remains low.

② When the sink capacity (e.g., number or size of panicles) is sufficiently high, photosynthesis in source leaves is enhanced and the rate increases.

③ However, if too many spikelets are formed (overdeveloped sink), the source may be unable to supply enough assimilates, resulting in lower grain filling rate and an increased proportion of unfilled grains (sterile spikelets).

2) Respiration

Plants utilize assimilates produced via photosynthesis to generate energy through respiration, which supports vital metabolic processes and the synthesis of secondary metabolites. Assimilates not consumed by respiration are stored. The energy required for plant life activities is derived from cellular respiration.

(1) Overview of Respiration

① Respiration is a process in which cells use oxygen (O_2) to oxidize glucose, producing energy and water (H_2O), while releasing carbon dioxide (CO_2) as a by-product.

② Glycolysis occurs in the cytoplasm, whereas aerobic respiration (including the TCA cycle and oxidative phosphorylation) takes place in the mitochondria. Alcoholic and lactic fermentation (anaerobic respiration) also occur in the cytoplasm.

③ During respiration, mitochondria use part of the released energy to synthesize ATP. Approximately 38% of the energy in one molecule of glucose (686 kcal) is stored as ATP.

④ In rice plants, ATP produced through respiration is used for the formation of new organs, nutrient uptake, and assimilate translocation.

(2) Respiration Rate

① In rice, respiration increases from transplanting through the establishment phase to the maximum tillering stage and gradually decreases thereafter.

② The amount of respiration per plant is generally proportional to photosynthesis and dry matter accumulation and typically peaks around the heading stage.

③ During the grain filling period, panicle respiration can account for up to 30% of the total plant respiration.

(3) Anaerobic Respiration (Fermentation)

① In rice, which germinates in water, seeds and roots often undergo anaerobic respiration (fermentation).

② Under submerged conditions with severe oxygen deficiency (e.g., flooding), rice roots can still perform anaerobic respiration.

③ While aerobic respiration yields 38 ATP per glucose molecule (2 from glycolysis, 2 from the TCA cycle, and 34 from oxidative phosphorylation), anaerobic respiration (alcoholic and lactic fermentation) produces only 2 ATP, resulting in poor energy efficiency and leading to chlorosis and tissue death.

Table 3-9. Comparison of aerobic and anaerobic respiration in rice

No.	Parameter	Aerobic respiration	Anaerobic respiration (Fermentation)	Notes
①	Use of oxygen	Utilized	Little or no use	Occurs under flooding/submergence
②	Substrate degradation	Complete breakdown into CO_2 and H_2O	Incomplete breakdown with intermediate products	Alcohol is a common by-product
③	ATP synthesis efficiency	38 ATP per glucose molecule	2 ATP per glucose molecule	Rapid depletion of energy
④	Site of occurrence	Cytoplasm and mitochondria	Cytoplasm only	Leads to fermentation

3) Relationship between Photosynthesis and Respiration

(1) Net Production (Photosynthesis - Respiration)

① To achieve maximum yield in crop production, net production, defined as the difference between photosynthetic assimilation and respiratory consumption, must be maximized. When photosynthetic output is high and respiration is low, the accumulation of organic matter increases. In rice, photosynthetic rates are approximately 5 to 10 times higher than respiration rates.

② The leaf in crops functions both as an assimilatory organ that conducts photosynthesis and as a respiratory organ that consumes energy.

③ Although increasing leaf area can enhance photosynthetic capacity, it also leads to higher respiratory consumption. Therefore, expanding leaf area does not necessarily result in increased net production or yield.

(2) Leaf Area Index (LAI)

① Leaf Area Index (LAI) is calculated as: Average leaf size × Number of leaves per tiller × Number of tillers per plant × Number of plants per unit area.

② The optimum leaf area refers to the leaf area at which net production is maximized.

③ LAI is the total leaf area of all rice plants grown within a given land area divided by the ground area.

④ The optimum LAI varies depending on growth stage, weather conditions, cultivar, and soil fertility.

⑤ Up to the maximum tillering stage, a higher LAI (up to 6~7) is advantageous, regardless of solar radiation.

⑥ The optimum LAI increases with higher solar radiation. When light intensity is low, the optimum LAI decreases, leading to reduced photosynthesis. Under high irradiance, rice photosynthesis increases almost linearly with LAI up to 6~7.

⑦ A larger optimum LAI for the rice canopy leads to greater photosynthetic production.

⑧ For standard temperate *Japonica* rice cultivars, the optimum LAI is typically around 5~6, while for Tongil-type cultivars, it is around 7~8.

⑨ Tongil-type cultivars generally exhibit a higher optimum LAI due to their erect leaf posture, which provides better light interception.

10. Transpiration and Water Requirement in Rice Cultivation

1) Transpiration

① In rice, as in most crops, greater transpiration is generally associated with higher yield.

② Rice utilizes only about 1% of the water it absorbs for metabolic processes such as photosynthesis and respiration, and the remaining 99% is released into the atmosphere primarily through stomata.

③ Transpiration plays a vital role in maintaining water balance, regulating plant temperature, and facilitating nutrient uptake.

④ Transpiration in rice occurs mainly through the leaf blade (lamina), though a portion also takes place via the awns.

2) Water Requirement

(1) Characteristics of Water Requirement

① Water requirement is defined as the amount of water needed to produce 1 g of dry matter. For transplanted lowland rice, the value is approximately 300, and for upland rice, about 400, which is lower than that of most other crops.

② Immediately after transplanting, rice roots are not yet fully functional, so deep irrigation is necessary to reduce transpiration and promote root establishment.

③ In rice cultivation, most water input serves environmental purposes. Physiological water demand peaks during the panicle initiation to heading stage, when reproductive growth is active, leaf area is large, and transpiration is high.

④ Water requirement is the reciprocal of water use efficiency (WUE), and it is equivalent to the transpiration coefficient.

⑤ Both water use efficiency and water requirement are dimensionless because the unit grams (g) cancel out. In the case of rice, the water requirement is around 300, which is relatively low among C_3 crops due to high yield potential, implying that rice has high water use efficiency.

Table 3-10. Water requirement in paddy rice cultivation (Kung, 1971)

<table>
<tr><th>Classification</th><th>Item</th><th>Water loss (mm/day)</th><th>Remarks</th></tr>
<tr><td rowspan="3">Water loss</td><td>Transpiration</td><td>1.5~9.8</td><td rowspan="4">Average from 43 locations in Korea, Japan, China, Thailand, Vietnam, and Bangladesh</td></tr>
<tr><td>Evaporation</td><td>1.0~6.2</td></tr>
<tr><td>Percolation</td><td>0.2~15.6</td></tr>
<tr><td colspan="2">Total</td><td>6~10</td></tr>
<tr><td rowspan="3">By growth stage</td><td>Nursery stage</td><td>40 mm/crop</td><td rowspan="4">Average 200 mm/month × 5 months = 1,000 mm + 500–600 mm for land preparation → approx. 8.64 mm/day = 8.64 m³/10a</td></tr>
<tr><td>Seedbed stage</td><td>200 mm/crop</td></tr>
<tr><td>Irrigation water use</td><td>1,000 mm/crop</td></tr>
<tr><td colspan="2">Total</td><td>1,240 mm/crop</td></tr>
</table>

(2) Water Use Efficiency (WUE)

① Water use efficiency refers to the effectiveness with which a crop utilizes the water it absorbs, calculated as the ratio of accumulated dry matter to water uptake during the growing period.

② WUE is typically expressed as the amount of dry matter produced (g) per unit of water consumed through evapotranspiration (g), i.e.,

WUE = Dry Matter Production (g) / Evapotranspiration (g)

③ WUE differs between individual plants and crop canopies, and it can be applied not only to total biomass production but also to economic yield.

④ WUE represents the efficiency of water use in relation to yield formation, reflecting how effectively water contributes to crop productivity.

⑤ WUE is inversely related to both the water requirement and the transpiration coefficient.

11. Yield Formation

1) Process of Yield Formation in Rice

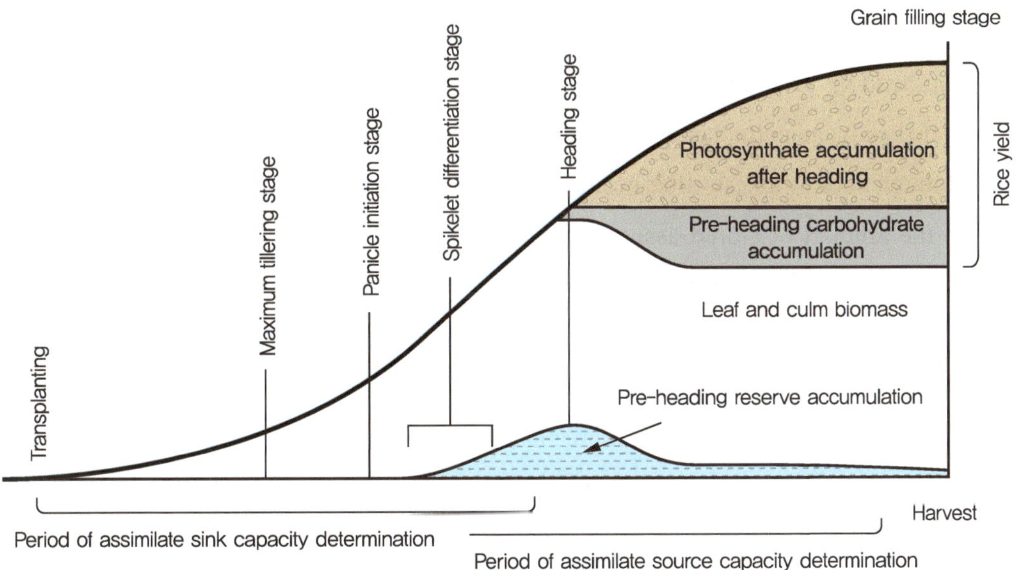

Figure 3-29. Growth stages and yield formation process in rice

(1) Yield Potential

① Yield potential is calculated as: Number of panicles per unit area × Number of spikelets per panicle × Volume (size) of individual spikelets.

② The factors determining yield potential are set during the approximately 80 days between transplanting and one week before heading, and are most influenced by the amount of nitrogen (N) applied as basal fertilizer and panicle fertilizer, as well as the amount of solar radiation received during this period.

③ When solar radiation during the ripening stage is abundant, greater yield potential results in increased yield. However, when solar radiation is limited, higher yield potential can lead to lower grain filling rate, which may reduce the actual yield.

(2) Yield Productivity

① Yield productivity is determined by the photosynthetic production system, the amount of assimilates produced, and the translocation of assimilates to the panicles.

② Key elements of the photosynthetic system include leaf area and the canopy structure (light interception capacity), with varietal selection playing the most critical role.

③ The final grain yield is the sum of the assimilates stored before heading and the photosynthates produced after heading.

④ Pre-heading accumulation includes approximately 30% of carbohydrates stored in the leaf sheath and culm that are translocated to the panicle, and 70% of assimilates produced through photosynthesis after heading.

⑤ If the amount of stored assimilates before and after heading is low, grain filling rate decreases and lodging becomes more likely. To increase these reserves, sufficient solar radiation and moderate vegetative growth (to reduce respiration) are required.

⑥ After heading, solar radiation becomes the most critical factor. To maintain high photosynthetic capacity and leaf area index during grain filling, root vitality is essential.

⑦ Regarding assimilate translocation to the panicle, the most influential factor is temperature. The optimum mean temperature for effective assimilate translocation in rice is 20~22°C. Below 17°C, translocation is inhibited, and excessive high temperatures increase respiration, thus reducing the amount of translocated assimilates.

⑧ Pre-heading accumulation begins approximately three weeks before heading, reaching its maximum at heading time, then rapidly decreases due to translocation to panicles.

⑨ Solar radiation exerts the greatest influence on rice yield production, with the most critical period being the seven weeks spanning three weeks before and four weeks after heading.

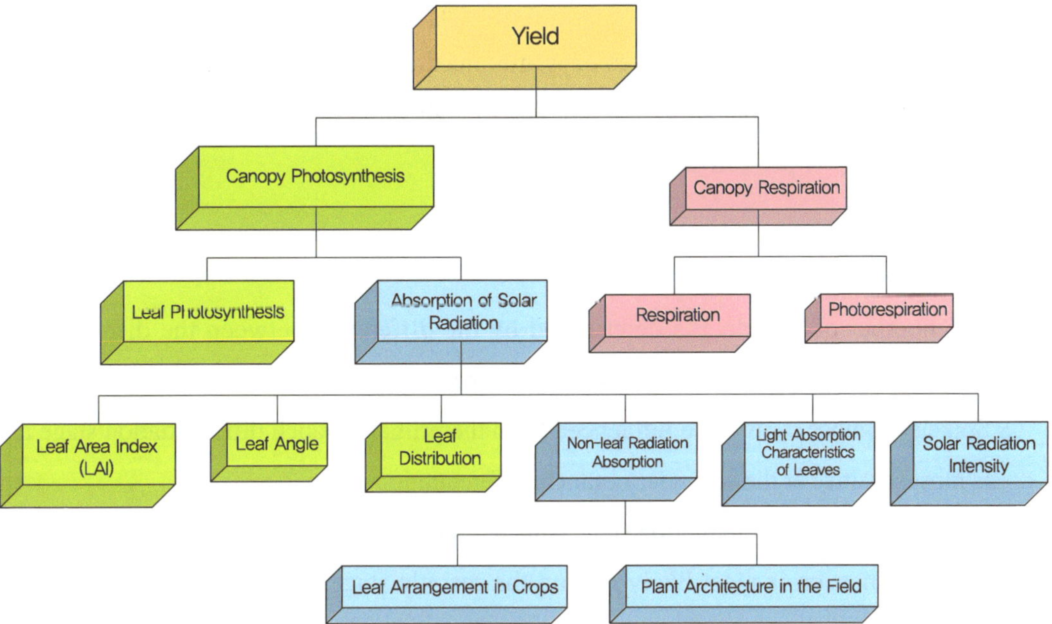

Figure 3-30. Flowchart of factors affecting rice yield

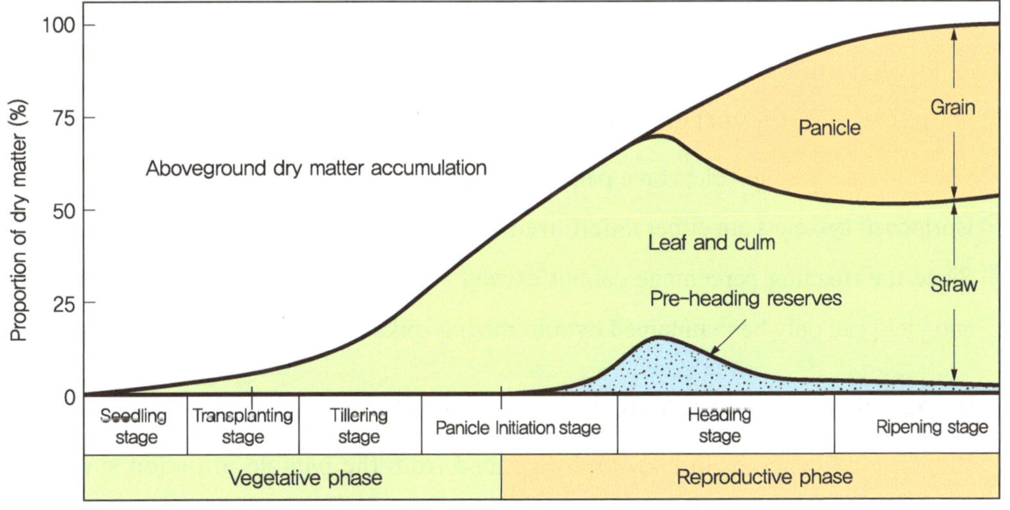

Figure 3-31. Dry matter accumulation in different growth stages of rice (Murata and Matsushima, 1975)

2) Yield Formation Process

(1) Panicle Number

① Although panicle number is partially influenced before transplanting, it is predominantly determined by post-transplanting environmental conditions.

② It is strongly affected during the active tillering stage, and little influenced after the spikelet differentiation stage (7~10 days after maximum tillering).

③ Panicle number is counted per hill, but due to high variability, counts are typically based on 10~30 hills or per 1 m^2 area.

(2) Number of Spikelets per Panicle

① The number of spikelets is determined by the difference between the differentiated spikelet primordia and the number of degenerated ones.

② Spikelet number starts to be influenced from the primary branch differentiation stage and is most sensitive at the secondary branch differentiation stage.

③ After the spikelet differentiation stage, spikelet number is rarely affected.

④ Spikelet abortion is most likely around the meiotic stage, and by about 5 days before heading (late meiosis stage), degeneration ceases and the spikelet number becomes virtually fixed.

(3) Percentage of Ripened Grains

A: Concept of ripening percentage

① The percentage of spikelets on a panicle that successfully develop into filled grains.

② Unripened spikelets are either unfertilized or aborted after fertilization.

③ Since the ripening percentage cannot exceed 100%, it does not directly increase yield, and yield can only be maintained by minimizing spikelet degeneration.

B: Influence of growth stages on ripening percentage

① Ripening percentage starts to be influenced from the panicle initiation stage and is especially vulnerable to decline during the meiosis stage, heading, and ripening phases.

② After about 35 days post-heading, it is scarcely affected.

③ Ripening percentage is usually calculated by sieving with a 1.85 mm screen or by using a salt solution of specific gravity 1.06 to separate filled grains.

C: Fertilization rate and temperature

① The fertilization rate is most sensitive to high temperature at heading, followed by approximately 9 days before heading.

② Exposure to high temperatures (~40°C) even for 1~2 hours during anthesis (flower opening) can significantly reduce fertilization.

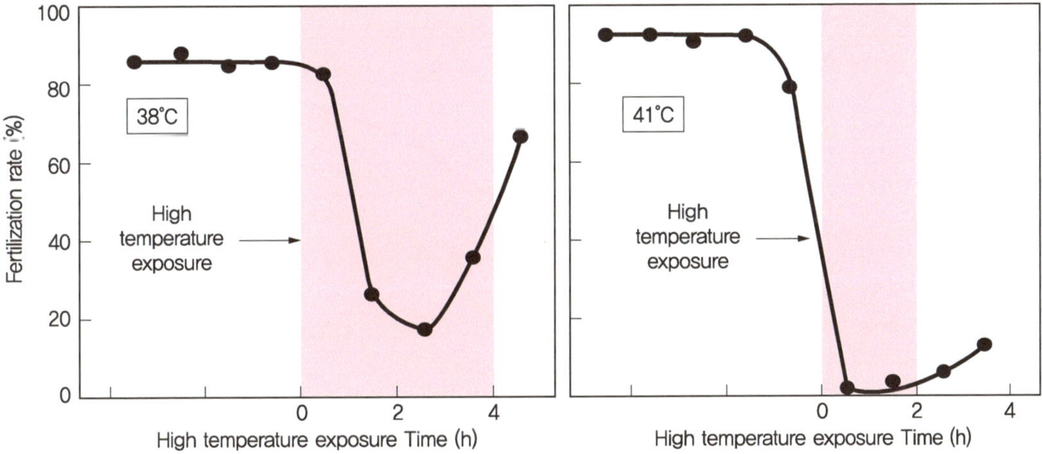

Figure 3-32. Effects of high temperature treatment (38°C and 41°C) on fertilization rate (Satake and Yoshida, 1978)

③ A 10-day exposure to low temperatures (~17°C) during meiosis can drastically reduce fertilization rate.

④ Application of nitrogen fertilizer tends to increase cold damage, whereas phosphate fertilizer can alleviate it.

(4) 1,000-Grain Weight

① The 1,000-grain weight is primarily determined by the size of the lemma and palea (hull) before heading. If the spikelet size is formed small, the brown rice grain will be mechanically restricted by the hull size regardless of favorable post-heading conditions.

② Secondarily, the 1,000-grain weight is influenced by the extent to which assimilates are filled into the spikelet after heading.

③ The critical period for promoting spikelet size enlargement is from the secondary branch differentiation stage through the spikelet differentiation and meiotic stages, and favorable environmental conditions during this period are important. After this stage, only the filling within the fixed spikelet size occurs, and it becomes difficult to significantly increase yield.

④ The most sensitive stages for reduction in 1,000-grain weight are the meiotic stage and the ripening stage.

⑤ The 1,000-grain weight is calculated by weighing approximately 23 grams of grain and counting the number of grains to determine the actual value.

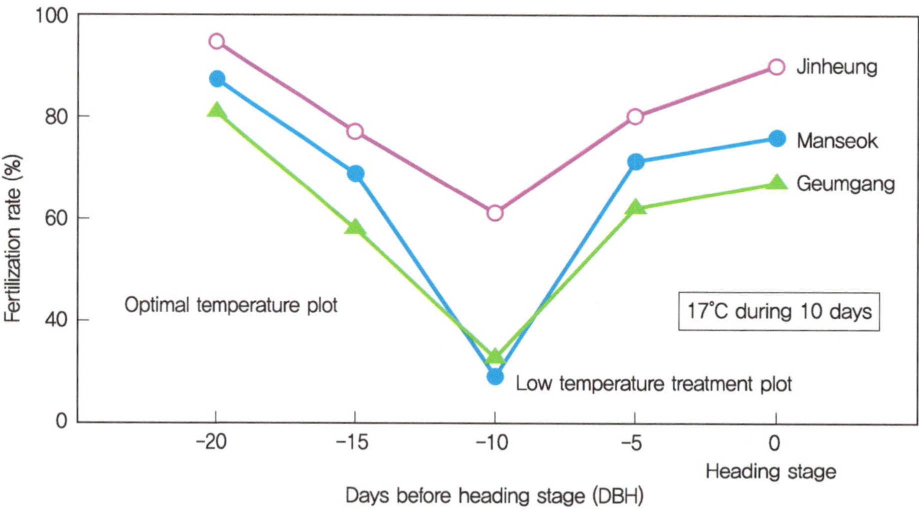

Figure 3-33. Relationship between low temperature (17°C for 10 Days) during meiosis and fertilization rate (Lee et al., 1979)

(5) Grain Yield

Grain yield is determined as the product of the four yield components. Until the floret differentiation stage, sufficient vegetative growth must be ensured to achieve the target yield, and after floret differentiation, efforts should focus on preventing any decline in the established yield potential.

A: Yield components and yield increase

① Increases in yield are largely influenced by two factors: the number of panicles per unit area and the number of spikelets per panicle.

② The first critical point for yield determination occurs at the peak tillering stage, which follows rapid increases in tiller number after transplanting. At this stage, yield increases through an increase in tiller and panicle numbers.

③ The second critical point appears at the stage of secondary branch primordium differentiation, where yield increases as a result of greater spikelet numbers per panicle.

④ After floret differentiation, there are essentially no further opportunities to actively increase yield.

B: Timing of yield increase and loss prevention

① Active yield enhancement is possible only up to the floret differentiation stage, and therefore the maximum yield potential is virtually determined by this stage.

② Yield loss prevention becomes critical after floret differentiation and is influenced by factors such as the degeneration of differentiated spikelets, the ripening of florets, and the extent to which assimilates are accumulated in the hull (lemma and palea), leading to grain filling and caryopsis enlargement.

Table 3-11. Effects of timed drought treatments before and after heading on yield components in rice (Matsushima, 1962)

No.	Timing of drought treatment relative to heading	Yield (g/hill)	Panicle number (per hill)	Sterility rate (%)	Ripening percentage (%)	1,000-Grain weight (g)
①	Control (no drought)	22.7	10	15	75	21.9
②	35 days before heading	20.0	12	11	60	20.5
③	27 days before heading	17.0	11	12	54	20.2
④	19 days before heading	15.7	11	34	52	20.8
⑤	11 days before heading	6.5	10	62	29	21.6
⑥	3 days before heading	8.3	10	59	38	20.9
⑦	5 days after heading	16.5	11	10	59	21.9
⑧	13 days after heading	20.5	10	7	66	22.5

3) Yield Components in Rice

(1) Number of Panicles

① The number of panicles per square meter (typically 400~500) is calculated as: Number of hills per m^2 × Number of panicles per hill.

② The number of hills per square meter is determined by planting density, usually 20~25 hills/m^2.

③ The number of panicles per hill is calculated as the average of 20 sample hills, typically ranging from 15 to 20.

(2) Spikelets per Panicle

① The number of spikelets per panicle is determined by counting the total number of spikelets on three representative hills (each with average panicle number), then dividing the total by the number of panicles.

② For temperate *japonica* cultivars, this value generally ranges from 80 to 100 spikelets.

(3) Grain Filling Rate (Fertility Percentage)

① The grain filling rate is calculated as follows: a sample of harvested and threshed panicles is randomly selected (about 30 g of grain), then immersed in saline water (specific gravity 1.06 for *japonica*, 1.03 for tongil types). The number of filled grains that sink is counted and divided by the total number of spikelets to determine the percentage.

② Typically, the grain filling rate is about 85% for *japonica* and 80% for tongil cultivars.

(4) 1000-Grain Weight (TGW)

① The 1000-grain weight is measured three times by weighing 1,000 grains and averaging the results.

② Since individual grain weights vary widely, 1,000 grains are weighed, and the total is divided by 1,000 to get the average per grain.

③ In recommended domestic cultivars, brown rice TGW is about 22 g, while milled rice TGW is around 20 g.

(5) Four Yield Components and Yield Calculation

① The four main components of yield are:

Number of panicles, Spikelets per panicle, Grain filling rate, and 1000-grain weight.

② Yield (Y) per unit area is calculated by combining panicle number and spikelet number per panicle, then multiplying by grain filling rate and 1000-grain weight.

> The formula is: Y (kg/10a) = PN × SN × F × W / 1000
>
> Where:
>
> PN = Panicles per m^2
>
> SN = Spikelets per panicle
>
> F = Grain filling rate (%)
>
> W = 1000-grain weight (g)

③ Alternatively, the yield per 10 ares can be calculated by multiplying the number of spikelets per square meter by the grain filling rate and the thousand-grain weight (TGW), and then dividing the result by 1,000.

> (Spikelets per m^2 × Grain filling rate × TGW) / 1000

④ In Korea, rice yield is expressed in kg per 10 a (1,000 m^2 = 1 danbo). Internationally, yield is typically expressed per 1 hectare (10,000 m^2).

⑤ Yield is measured as milled rice (baekmi) in Korea, brown rice (genmai) in Japan, and rough rice (paddy) globally.

⑥ In calculating rice yield, it is important to note that 10 ares is equivalent to 1,000 square meters or 300 pyeong. The grain filling rate, expressed as a percentage, must be divided by 100 to convert it to a decimal. The thousand-grain weight (TGW), given in grams per 1,000 grains, should be converted to kilograms by dividing by 1,000. Based on these conversions, the yield can be calculated using the formula: spikelets per square meter multiplied by the grain filling rate (as a decimal) multiplied by the TGW (in kilograms), or simplified as panicle number (PN) times spikelet number per panicle (SN) times grain filling rate (F) times grain weight (W), all divided by 100,000.

For example, if there are 30,000 spikelets per square meter, the grain filling rate is 80 percent, and the TGW is 20 grams, then the estimated yield would be calculated as 30,000 multiplied by 80 multiplied by 20, divided by 100,000, resulting in a yield of 480 kilograms per 10 ares.

> In calculating yield:
> 10a = 1,000m² = 300 pyeong
> Grain filling rate is in percent (%) = divided by 100
> TGW is per 1,000 grains, in grams = converted to kg by dividing by 1,000
> Thus, yield = Spikelets/m² × (1/1,000) × Grain filling rate (1/100) × TGW (g) × (1/1,000)
> = PN × SN × F × W / 100,000
> Example: If spikelets/m² = 30,000, grain filling rate = 80%, and TGW = 20g: → Yield
> = 30,000 × 80 × 20 / 100,000 = 480kg/10a

⑦ In a rice field with a planting density of 24 hills per square meter and four seedlings transplanted per hill, each seedling produces an average of five tillers. Assuming there are 100 spikelets per panicle, a grain filling rate of 80 percent, a 1,000-grain weight of brown rice of 22 grams, and a conversion factor from brown rice to paddy rice of 1.25, the estimated number of panicles per square meter is 480. Based on this, the grain yield per square meter is calculated to be 844.8 grams. This corresponds to a yield of 844.8 kilograms per 10 ares. After converting to paddy yield, the final estimate is approximately 1,056 kilograms per 10 ares.

> Tillers per seedling = 5
> Spikelets per panicle = 100
> Grain filling rate = 80%
> Brown rice TGW = 22g
> Conversion factor from brown rice to paddy = 1.25
> → Panicles per m² = 24 hills × 4 seedlings × 5 tillers = 480 panicles
> → Grain yield per m² = 480 × 100 spikelets × 0.8 × 0.022g = 844.8g/m²
> → Yield per 10 ares = 844.8 × 1,000 = 844.8kg/10a
> → Paddy yield per 10 ares = 845 × 1.25 = 1,056kg/10a

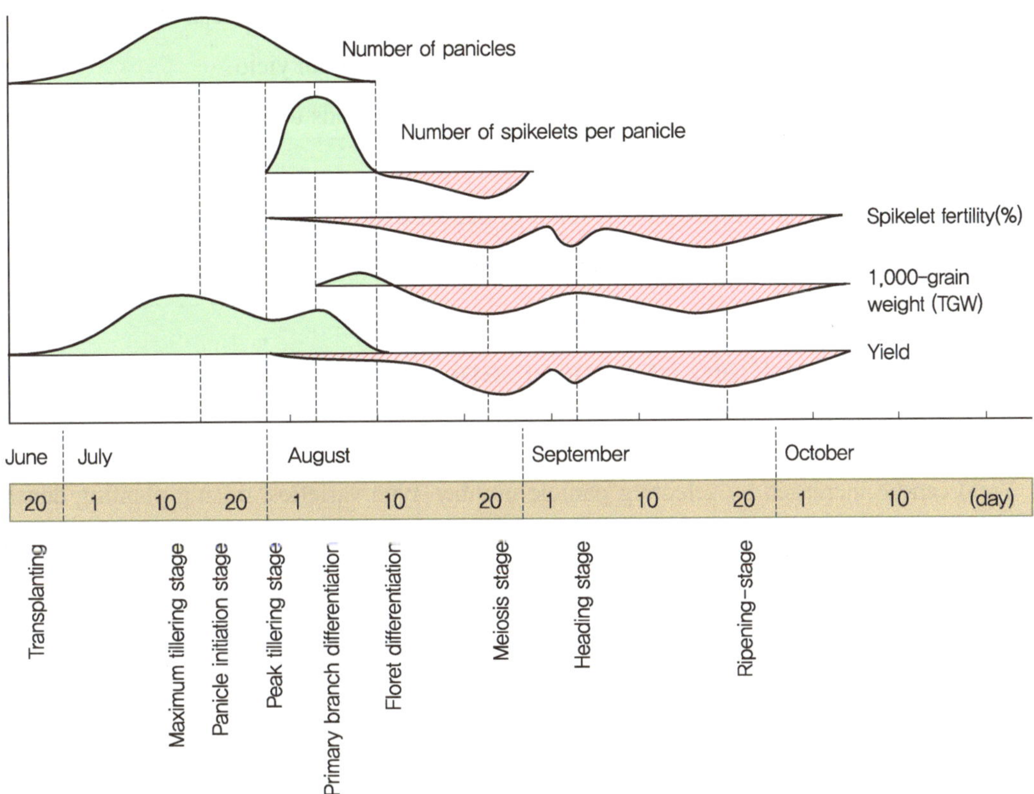

Figure 3-34. Schematic diagram of yield formation in rice

(6) Coefficient of Variation (CV) of Yield Components

The order of annual variation in the coefficient of variation among yield components is: grain yield, panicle number per unit area, number of spikelets per panicle, ripening ratio, and 1,000-grain weight. Generally, the coefficient of variation is approximately 14% for grain yield, 12% for panicle number, 10% for spikelets per panicle, 7% for ripening ratio, and 3% for the 1,000-grain weight of brown rice.

(7) Complementarity of Yield Components

① The yield components, ranked in order of their influence on grain yield, are panicle number per unit area, spikelets per panicle, ripening ratio, and 1,000-grain weight.

② Complementarity refers to the interrelated and compensatory relationships among the four yield components. When an earlier-forming component is produced in excess, subsequent components tend to be reduced. Conversely, if an earlier component is reduced, the later ones may increase to compensate, thereby stabilizing the total yield.

③ In general, an increase in panicle number per unit area leads to a reduction in the number of spikelets per panicle.

④ An increase in spikelets per panicle tends to decrease the ripening ratio.

⑤ A decrease in ripening ratio, in turn, leads to an increase in 1,000-grain weight.

(8) Strategies for Yield Enhancement

A: Securing panicle number

Yield can be increased by selecting panicle number–type varieties, raising planting density through close or early planting or multiple seedlings per hill, and applying high basal and tillering fertilizers. Cultivation practices that enhance tillering, such as shallow irrigation and adequate nitrogen management, are essential. Deep transplanting should be avoided.

B: Increasing spikelets per panicle

Use varieties with short culms and heavy panicles (e.g., Tongil-type rice), and apply heading-stage nitrogen fertilizer (panicle fertilizer).

C: Improving ripening ratio

Control panicle number and spikelets per panicle to maintain an optimal number of spikelets per unit area. Transplant at an appropriate time to ensure heading occurs before the threshold date for safe ripening.

Suppress ineffective tillers and avoid delayed effective tillering. Apply grain-filling fertilizer to increase grain weight. For high yields, even with high panicle or spikelet numbers, a ripening ratio of 75~80% is desirable. Values below this suggest excessive spikelet formation.

4) Harvest and Yield

(1) Harvest Index (HI)

① The harvest index is the ratio of economically usable biomass (i.e., dry grain weight) to total aboveground biomass.

② It represents the proportion of economic yield (grain yield) to biological yield (total biomass).

③ Typical harvest indices are approximately 0.30 for traditional tall varieties, 0.50 for temperate *japonica* semi-dwarf varieties, and 0.55 for Tongil-type semi-dwarf varieties.

$$\text{Harvest index (HI)} = \frac{\text{Economic yield}}{\text{Biological yield}} = \frac{\text{Dry grain weight}}{\text{Total aboveground biomass}}$$

(2) Grain-Straw Ratio (G/S Ratio)

① The grain-to-straw ratio represents the weight ratio of rough rice (unhulled grain) to rice straw and is calculated by dividing the weight of rough rice by the dry matter weight of the straw.

② The grain-to-straw ratio is approximately 0.4 to 0.5 in traditional tall rice varieties, around 1.0 in temperate *japonica* semi-dwarf varieties, and about 1.2 to 1.3 in Tongil-type short-culm, heavy-panicle varieties.

$$\text{Grain-straw ratio} = \frac{\text{Dry grain yield}}{\text{Dry straw weight}}$$

(3) Yield and Indicators

① Grain yield can be enhanced by increasing either the total dry matter production (an indicator of canopy photosynthesis in the field) or by increasing the harvest index.

② The harvest index indicates the proportion of biologically produced dry matter that is economically harvestable.

③ For typical rice cultivation, total aboveground biomass ranges from 1,000 to 1,500 kg per 10 ares, and grain yield ranges from 400 to 800 kg per 10 ares under optimal conditions without lodging, pest outbreaks, or weather damage.

(4) Yield Estimation

A: Yield survey of rice

① The most accurate method to determine rice yield is to harvest the entire field and weigh the total yield. However, this is impractical in most situations. Therefore, yield is generally estimated using samples taken from selected plots based on statistical sampling methods.

② Yield components are estimated using methods such as the visual estimation method, spikelet counting method, and small-area harvesting method.

③ The current national record for rice yield in Korea is 1,006 kg/10a using the variety Sankangbyeo. Global yield records also range between 1,000~1,100 kg/10a. For forage rice, yield is often evaluated by total biomass, including straw. However, rough rice yield alone can sometimes reach 1,200 kg/10a (refer to Table 4-4).

B: Photosynthetic efficiency

① The theoretical maximum photosynthetic efficiency of higher plant leaves is approximately 30% of incoming solar radiation at the earth's surface.

② Under ideal conditions, the photosynthetic efficiency of crop canopies is around 20%.

③ When grain yield reaches 1,000 kg per 10a, the conversion efficiency of solar energy into biomass is about 3%.

④ In actual farm conditions, energy conversion efficiency typically ranges from 1.6% to 2.0%.

⑤ Dense planting at optimal spacing generally increases canopy light interception and photosynthetic efficiency, thereby enhancing grain yield (see Table 3-12).

Table 3-12. Variation in rice yield according to planting spacing (Mabbayad and Obordo, 1970)

No.	Spacing (cm)	Yield (kg/10a)		
		Low-tillering variety*	High-tillering variety	Traditional high-tillering variety
①	15 × 15	620	760	410
②	25 × 25	570	820	460
③	35 × 35	520	740	440
④	45 × 45	420	690	430
⑤	55 × 55	350	650	360

*Low-tillering varieties produce fewer tillers, while high-tillering varieties produce more tillers.

Rice Crop Science

Chapter 4. **Rice Varieties and Genetic Breeding**

Chapter 4.
Rice Varieties and Genetic Breeding

The Tongil-type rice varieties opened the era of rice self-sufficiency in Korea. The development of Tongil-type cultivars was made possible through the effective combination of classical breeding theory and international collaboration, including support from the International Rice Research Institute (IRRI). However, due to their susceptibility to low temperatures and unsatisfactory eating quality, these varieties have now nearly disappeared. This trend reflects a historical shift in breeding priorities from yield to quality, underscoring the continued need for breeding efforts.

Despite their decline in domestic cultivation, Tongil-type rice continues to exert a positive influence globally, being widely used as a genetic resource (e.g., as donor parents), particularly in China. Undoubtedly, such breeding successes would not have been possible without the accumulated theoretical foundation of genetics and breeding, as well as advances in crop physiology and agronomic practices. Although there has been ongoing debate over whether yield improvements are primarily due to agronomic techniques or genetic breeding, it is increasingly recognized that breeding plays an essential role. In this chapter, we examine several major rice varieties, their characteristics, and directions for future improvement.

1. Characteristics of Rice Varieties

1) Classification of Rice

(1) Classification by Ecological Characteristics

① Based on ecological differentiation, rice is classified into temperate *japonica*, tropical *japonica* (*javanica*), and *indica* types.

② Temperate *japonica* varieties have short, round grains that are not easily shed. Their leaves are dark green, plant height is short, and seedlings show strong cold tolerance but weak drought resistance.

③ *Indica* varieties have long, slender grains that are easily shed. Their leaves are light green, plants are tall, and seedlings exhibit weak cold tolerance but strong drought resistance.

④ Depending on the cultivation environment, rice can be classified as paddy rice (cultivated in flooded fields) and upland rice (grown in dry fields).

⑤ Based on growth duration from sowing to maturity, rice is categorized into early-maturing, medium-maturing, medium-late-maturing, and late-maturing types.

(2) Classification by Morphological Traits

① Plant height depends primarily on internode length, with classifications such as tall and semi-dwarf types. Extremely short varieties are referred to as dwarf types.

② Grain shape is classified as short grain, medium grain, or long grain based on grain length and width.

Table 4-1. Comparative characteristics of cultivated rice types

No.	Trait category	Trait	*Indica*	Temperate *Japonica*	Tropical *Japonica*
①	Cooking quality	Amylose content	25%	17~20%	20%
		Chewiness	Hard	Soft	–
		Stickiness	Low	High	Moderate
②	Distribution	Geographic distribution	India, Vietnam, Thailand	Korea, Japan, Northern China	Java (Indonesia)
③	Physiological	Cold tolerance	Weak	Strong	Moderate to strong
		Drought resistance	Strong	Weak	Weak to moderate
④	Plant traits	Leaf color	Light green	Dark green	Light green
		Plant height	Tall	Short	Tall
		Plant type	Erect	Spreading	–
		Stem hardness	High	Low	High
		Tillering ability	High	Moderate	Low
		Tillering angle	Spreading type	Closed type	Closed type
⑤	Seed traits	Awn presence	Absent	Present (absent in bred types)	Variable
		Grain shape	Long and slender	Short and round	Long and slender
		Shattering	High	Low	Variable

③ Temperate *japonica* varieties cultivated in Korea have a brown rice length-width ratio of approximately 0.6 to 2.1, which is classified as short grain.

④ Based on grain size, rice is divided into large, medium, and small grain types. Most table rice temperate *japonica* varieties fall under small grain, with a thousand-grain weight of approximately 22g.

⑤ For processing purposes, some temperate *japonica* varieties have large grains with a thousand-grain weight of about 35g.

⑥ Depending on the presence or absence of awns, rice is classified as awned or awnless varieties.

(3) Classification by Chemical Composition

A: Classification based on starch type

① Non-glutinous rice (ordinary rice) contains approximately 20% amylose and 80% amylopectin as storage starch components. High-quality table rice with better eating quality typically contains a lower amylose content, ranging from 15% to 18%.

② Glutinous rice (sticky rice) contains more than 90% amylopectin, resulting in a sticky texture and a high degree of gelatinization even at the core of the grain.

③ Amylose is structurally more accessible to the digestive enzyme amylase, making non-glutinous rice more digestible than glutinous rice.

④ Glutinous rice appears milky white and opaque because its starch granules contain micro-voids that cause diffuse light reflection.

⑤ In general, glutinous rice yields are lower than those of non-glutinous rice, but the purchase price is typically higher.

⑥ The distinction between non-glutinous and glutinous rice can be made using the iodine staining method: non-glutinous rice stains blue-black, while glutinous rice stains reddish-brown.

⑦ The amylose content of rice is regulated by the genotype of the Waxy (Wx) gene located on chromosome 6, and rice is classified accordingly into glutinous rice (8% amylose), low-amylose rice or partially glutinous rice (13%), ordinary rice (20%), and high-amylose rice (27%).

B: Aroma and pericarp characteristics

① Based on the intensity of aroma, rice can be classified into scented rice and highly scented rice.

② Based on pericarp color, rice is classified into white rice and pigmented rice, with the latter, also referred to as jadobyeo in Korean, further subdivided into black rice and red rice.

③ Although scented rice and milky rice (white-core rice) generally have lower yields, they are preferred by certain consumer groups due to their distinctive traits.

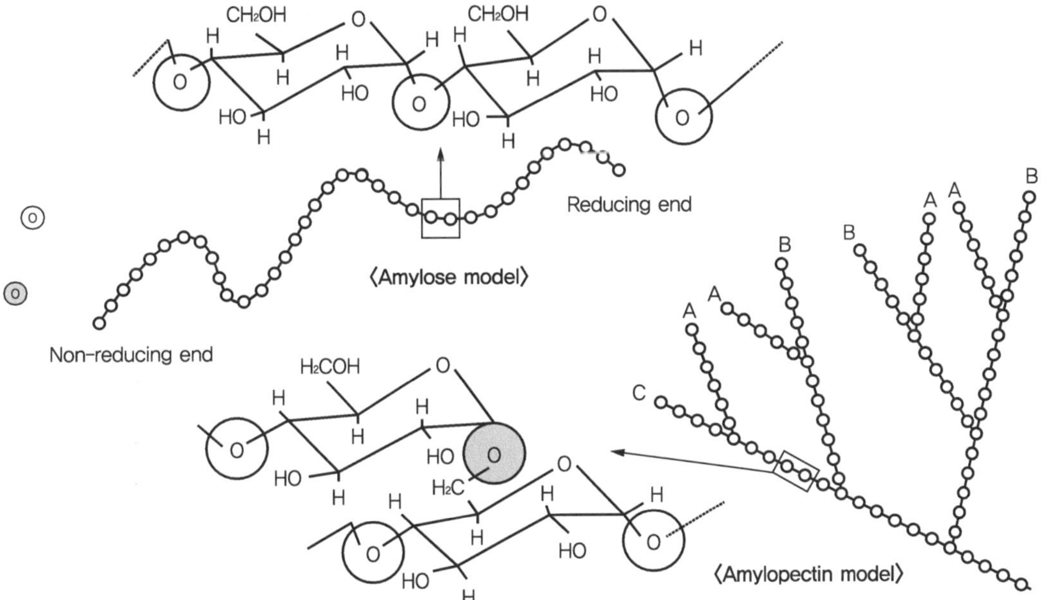

Figure 4-1. Schematic diagram of amylose and amylopectin (○: Glucose unit, ◎ : α-1,4 linkage, ◉ : α-1,6 linkage)

2) Evolution of Rice Varieties

(1) Chronological Stages of Variety Development

A: Era of landraces (Before 1915)

① This period refers to a time when 100% of rice cultivated in Korea consisted of traditional landraces.

② Major landraces included Noinjo, Dadajo, Maekjo, and Jodongji.

③ These varieties were characterized by low yield and high susceptibility to lodging.

B: Era of introduced vrieties (1916–1935)

① This was a period when introduced varieties from Japan rapidly replaced local landraces, becoming mainstream. The proportion of introduced varieties rose from 2% in 1912 to 60% in 1920, 74% in 1927, and over 80% after 1933.

② Representative varieties included Gokryangdo, Josinryeok, Tamagum, and Eunbangju.

C: Era of domestic breeding and dissemination (1936–1970)

① Korea's first artificial hybridization was performed in 1914, leading to the dissemination of the first bred cultivars, Namsun No.13 and Pungok, in 1933.

② This was a period when domestic varieties and introduced ones were cultivated in roughly equal proportions. Major domestic varieties included Paldal and Palgwang, while Eunbangju and Nonglim No.6 were prominent introduced cultivars.

D: Era of tongil-type varieties (1971–1980)

① Tongil rice was developed by the Rural Development Administration (RDA) and was featured on the 50-won coin first issued in 1972.

② Since its official registration as a recommended variety in 1971, Tongil rice spread rapidly, and by 1976 Korea achieved self-sufficiency in rice. Initially, a rice panicle (ear) of Tongil rice contained 28 spikelets, but this increased to 43 from 1983 onward, with smaller and more realistic florets (typically, one panicle bears around 90 florets).

③ Tongil rice was derived from the crossing of IR8 × (Yukara × Taichung Native 1). IR8 and TN1 are *indica* types, while Yukara is *japonica* type, making it a three-way *indica–japonica* hybrid.

④ As the first variety developed through a three-way hybridization involving distantly related germplasm, Tongil rice had a semi-dwarf, high-yielding, fertilizer-responsive, and disease-resistant plant type.

⑤ The variety possessed semi-dwarfing genes, conferring high fertilizer tolerance and lodging resistance, but it was vulnerable to low temperatures, leading to severe cold damage in 1980.

⑥ While Tongil rice was resistant to neck blast (although resistance weakened again in 1978) and stripe virus disease, it was susceptible to bacterial leaf blight and rice stem borers.

Table 4-2. Hybridization schemes in rice breeding

No.	Category	Breeding method description
①	Single cross	A simple cross between two genetically distinct varieties (e.g., A/B). Most rice cultivars bred in Korea use this method.
②	Three-way cross	A hybrid of F_1(A/B) crossed with a third parent C (e.g., A/B//C or C//A/B).
③	Multiple cross	Crossing an F_1 hybrid with one of its parents (e.g., A/B//A, A/A//B, or A/B//B). Used to improve a single trait from an existing cultivar. Typically involves 4~6 backcross generations. The donor used only once is the pollen parent (non-recurrent parent), while the repeatedly used parent is the seed parent (recurrent parent).
④	Backcross	A complex hybrid involving multiple cultivars (e.g., A/B//C/D//E/F//G), often used to develop near-isogenic lines (NILs)–lines genetically uniform except for one or a few target genes. Effective for developing varieties with polygenic resistance.

*For example, the proportion of A's genome in A/B//C is 25%, in A/B//C/D is 25%, and in A/B//A///C is 37.5%.

⑦ Although Tongil rice initially had a major drawback of high shattering, this was later improved. As a result, the cultivation area expanded to 76% by 1978. However, in 1980, a severe cold spell drastically reduced yield.

⑧ Tongil rice had low stickiness and poor eating quality, leading to its discontinuation in recent years. Nevertheless, it made significant contributions as a donor parent in breeding programs, both domestically and internationally.

⑨ To overcome its weak cold tolerance, a cultivation technique known as protected seedling raising (thermal seedling method) was developed.

⑩ Tongil rice also suffered from low ripening ratio and low milling recovery rate.

⑪ Major Tongil-type cultivars included Tongil, Yushin, Milyang 21, and Milyang 23, the latter of which once held the world record for highest yield.

E: Decline of the tongil-type rice era (1980–1990)

① Due to economic growth and an increasing preference for better eating quality, especially palatability, the cultivation area of Tongil rice dropped to 50% by 1980 and further declined to 19% by 1987.

② Major Tongil-type cultivars during this period included Yushin, Geumgangbyeo, Milyang 21, Milyang 23, and Milyang 30. In contrast, temperate *japonica* cultivars that gained popularity during this time included Nakdongbyeo and Akitakomachi (Chucheong).

F: Era of high-quality and functional varieties (1990–present)

① This period saw the widespread adoption of premium-quality rice varieties and the development and dissemination of multi-purpose rice. Tongil rice cultivation dropped to 4% by 1991, and by 1992, it had almost disappeared from farmer fields.

② Representative early-maturing high-quality varieties include Odaebyeo, Jinbubyeo, and Jinmibyeo, medium-maturing varieties include Ilpumbyeo, Jang-anbyeo, and Hwaseongbyeo, and medium-late-maturing varieties include Dongjinbyeo and Saechucheongbyeo.

③ Among functional rice varieties, Yangjobyeo is used for brewing, Daeanbyeo for direct seeding, Obongbyeo is a giant embryo rice, Hyangmi 1 and Hyangnambyeo are aromatic varieties, Daelib 1 is a large-grain variety for processing, and Dasanbyeo, Namcheonbyeo, Andabyeo, Areumbyeo, and Hanmaeum are ultra-high-yielding varieties derived from improved Tongil-type rice.

④ Tongil-chalbyeo (glutinous Tongil-type rice) was developed through backcross breeding, while Baekjinju and Seolgaengbyeo, low-amylose (9%) varieties, were selected from irradiated Ilpumbyeo zygotes using the mutagen NMU (N-nitroso-N-methylurea) in 2002.

⑤ With the release of Nokyangbyeo in 2006, the forage rice era began. Subsequent varieties such as Mokwoo, Youngwoo, and Chungwoo have shown total biomass yields over 2,000 kg per 10a.

⑥ Top Rice is a premium eco-friendly rice brand that meets strict production and quality standards, including collective cultivation with traceability systems, prevention of varietal admixture, timely transplanting with optimal hill density, eco-friendly practices to minimize pest and disease damage, supply based on rigorous quality inspection criteria, cold storage to maintain rice quality with more than 95% fully milled grains, soil and water analysis to apply only the necessary amount of fertilizers, and reduced protein content to enhance eating quality

Figure 4-2. Appearance of the ‘Top Rice’ packaging bag

Table 4-3. Functional rice varieties and processing suitability (Source: 2021 Rural Development Administration Website, Korea)

No.	Purpose	No. of varieties	Major varieties	Processing suitability and characteristics
①	High yield	20	Anda, Areum, Boramchan, Dasan, Dasan1, Dasan2, Durechan, Geumgang1, Hanareum, Hanareum2, Hanareum3, Hanareum4, Hanmaeum, Himangchan, Keunseom, Namcheon, Namil, Segyejinmi	• Tongil-type super high yield (Geumgang 1: 817kg/10a) • *Japonica*-type super high yield(Boramchan: 733kg/10a)
②	High-amylose	7	Dodamssal, Goami, Mimyeon, Palbangmi, SaeGoami, SaeMimyeon, Saeromi	• High amylose content, suitable for rice noodles, rice gel
③	Low-amylose (Waxy)	24	Baekjinju No.1, Baekjinju, Baekokchal, Baekseolchal, Boramchal, Boseokchal, Cheongbaekchal, Dongjinchal, Haephyeongchal, Hanareumchal, Hangangchal1, Hwaseonchal, Jinbuchal, Jinseolchal, Manmi, Miho, Mireuchal,Noonbora, Sangjuchal, Seolbaek, Sinsunchal, Unbaekchal, Unilchal, Wolbaek	• Glutinous rice: virtually no amylose, high stickiness – Traditional foods, rice cakes, Korean confections • Intermediate waxy rice(9–13%amylose): kimbap, brownrice(fordiabetics) • Tongil-type glutinous rice: high-yield waxy type

④	Pigmented rice	19	Boseokheukchal, Cheonghyangheukmi, Dahongmi, Geonganghongmi, Heukgwang, Heukjinju, Heukjinmi, Heuknam, Heukseol, Heuksujeong, Hongjinju, Jeokjinju, Jinheukchal,Joeunheukmi, Josaengheukchal, Seonhyangheukmi	• Black and red rice types • Mixed with brown rice for consumption • Anti-aging, antioxidant properties • Natural colorants (cosmetics, snacks, etc.)
⑤	Aromatic rice	9	Arome, Aryanghyangchal, Durehyang, Hyangnam, Hyangmibyeo1, Hyangmibyeo2, Hyangyeol, Mihyang, Seolhyangchal	• Enhanced palatability via aroma • For mixed rice, rice cakes, traditional sweet drinks (sikhye), etc.
⑥	Functional rice	1	Yeonganbyeo	• Child growth, high lysine (4%)
		1	Goami4	• Rich in iron and zinc (41~67%), it helps prevent anemia
		2	Goami2, Goami3	• Diet-oriented, High in resistant starch • Not suitable for table rice
		4	Keunnun, Keunpum, Noonkeunheukchal, Noonkeunheukchal1	• Large embryo (3×), GABA-rich • Germinated rice, blood pressure, Suitable for mixed grain rice
		2	Geonyangmi, Geonyang2	• Kidney disease diets, low-glutelin
⑦	Rice flour varieties	4	Baromi, Hangaru, Misiru, Singil	• Soft-textured varieties for rice flour • Suitable for dry and wet milling
⑧	Brewing rice	2	Seolgaeng, Yangjo	• Soft grains, porous, suitable for brewing
⑨	Frying & brewing rice	1	Daeripbyeo1	• Large-grain (1.5× size)
⑩	Sugary rice	1	Danmi	• Naturally sweet rice, for snacks and beverages
⑪	Brown rice	1	Bodurami	• Easy gelatinization, soft texture, suitable for brown rice
⑫	Upland rice	2	Nongrimna1, Sangnambatbyeo	• Suitable for upland cultivation, glutinous type
⑬	Forage rice	10	Cheongwoo, Gowoo, Jonong, Jowoo, Miwoo, Mokwoo, Mokyang, Nokwoo, Nokyang, Youngwoo	• Silage use, high biomass • Mokwoo: 2,059kg/10a total biomass

G: Forage rice varieties

① Recently, the demand for forage rice has been increasing in countries such as Korea and Japan, leading to active breeding of rice varieties for this purpose.

② Forage rice can be classified into grain-type forage rice, whole crop silage (WCS), and rice straw.

Table 4-4. Varieties and characteristics of forage rice in Korea

No.	Variety name	Year of development	Total Forage dry matter yield (kg/10a)	Main characteristics
①	Nokyang	2006	1,652	Medium–late maturing, persistent late-season greenness, shattering-resistant, lodging-resistant, resistant to bacterial leaf blight
②	Moku	2009	2,059	Late maturing, persistent late-season greenness, lodging-resistant, multi-disease resistance
③	Mokyang	2010	1,770	Medium–late maturing, lodging-resistant, resistant to virus diseases (stripe disease, rice black streaked dwarf disease)
④	Noku	2014	1,646	Medium–late maturing, good total dry matter yield, cold-germination tolerance, good early growth vigor
⑤	Youngu	2015	2,000	Medium–late maturing, high total dry matter yield, resistance to multiple diseases and pests, cold-germination tolerance, good grain filling rate
⑥	Chengu	2016	2,057	Late maturing, high total dry matter yield, resistance to multiple diseases and pests, lodging-resistant, cold-germination tolerance
⑦	Miwoo	2017	1,988	Medium–late maturing, high total dry matter yield, excellent forage quality, cold-germination tolerance, high amylose content
⑧	Jowoo	2018	1,817	Early–medium maturing, cold-tolerant, lodging-resistant, moderate shattering tendency, resistant to rice stripe virus
⑨	Gowoo	2019	1,820	Medium–late maturing, smooth leaves and grains without trichomes, lodging-resistant

Table 4-5. Trends in forage rice cultivation area in japan

Item	1995	2000	2006	2010	2016	2022
Area (ha)	23	502	5,182	15,939	91,000	142,000

(2) Status of Rice Variety Development National Seed Management Office List

① Rice varieties in Korea are registered in the National Variety List and can be accessed through the National Seed Management Office website (www.seed.go.kr).

② The varieties are diverse in terms of maturity group and intended use, including high-quality rice, varieties adapted to specific regions, specialty rice for processing, ultra-high-yielding varieties, and upland rice.

Table 4-6. Characteristics of major rice varieties

No.	Variety name	Year of development	Intended use	Yield (kg/10a)	Characteristics
①	Cheongpum	2015	Table rice (premium quality)	536	Resistant to rice blast, bacterial leaf blight, and rice stripe virus
②	Chucheong	1969	Table rice	547	This medium–late maturing variety is lodging-resistant, susceptible to rice blast, rice black streaked dwarf virus, and brown planthopper, and is characterized by good grain appearance, excellent eating quality (protein 6.6%), and high milling yield
③	Damageum	1908	Table rice	135 (estimated)	Farmer-selected
④	Dasan	1995	Processing use (Tongil-type)	677	Resistant to rice blast, bacterial leaf blight, and rice stripe virus
⑤	Heukjinju	1998	Eating quality improvement (black-purple rice)	405	This extremely early maturing variety, developed through anther culture, has poor cold germination ability and high anthocyanin content
⑥	Hiami	2008	Table rice	538	High content of 8 essential amino acids (131% compared to Hwaseong)
⑦	Hwaseong	1985	Table rice (high quality)	493	This erect plant type variety, developed through anther culture (haploid breeding), is resistant to rice stripe virus

⑧	Hyangmi 1	1993	Eating quality improvement	493	This Tongil-type variety shows resistance to rice stripe virus
⑨	Ilpum (Gyeongbuk)	1990	Table rice (high quality)	535	This medium–late maturing variety is high quality with a high embryo rice ratio, susceptible to rice blast, sheath blight, and brown planthopper, and shows an erect plant type with lodging resistance
⑩	Jodongji (landrace)	Landrace	Table rice	135 (estimated)	Selected from farms in Yeoju in 1988
⑪	Jungsaeng Eunbangju	1938	Table rice	200 (estimated)	Medium-maturing variety selected and purified from Eunbangju
⑫	Milyang 23	1976	Table rice (Tongil-type, high quality)	576	This variety is erect in type, resistant to lodging and rice stripe virus, and produces clear grains
⑬	Nakdong	1975	Table rice (high quality)	498	It is resistant to saltwater damage and resistant to rice stripe virus
⑭	Nokyang	1991	Forage use (whole crop)	1,652	This variety maintains late-season greenness, resists shattering and lodging, and is resistant to bacterial leaf blight and rice stripe virus, though susceptible to rice blast
⑮	Nong-an	1993	Table rice (direct seeding)	503 (transplanted) / 476 (dry direct seeding)	This variety produces few tillers, shows an erect growth habit, and resists lodging
⑯	Odae (Gangwon)	1983	Table rice	460	This extremely early maturing variety shows strong cold tolerance and cold germination ability but is susceptible to pests, diseases, and premature heading
⑰	Paldal	1944	Table rice	300	Strong cold tolerance
⑱	Samgwang (Chungnam)	2003	Table rice (premium quality)	569	This medium–late maturing variety is susceptible to brown planthopper and rice black streaked dwarf virus but resistant to rice blast, bacterial leaf blight, and rice stripe virus

⑲	Sindongjin (Jeonbuk)	1999	Table rice	596	This medium–late maturing variety has medium–large grains, strong cold tolerance, susceptibility to rice blast and brown planthopper, and high yield potential.
⑳	Tongil	1971	Table rice	449	This variety produces a medium number of tillers, grows erect, and resists lodging and bacterial leaf blight

*Variety names in parentheses indicate the main cultivation region. As of 2022, 'Sindongjin' (Jeonbuk) is the most widely cultivated variety nationwide.

(3) Seed Distribution

A: Seed distribution system

① In Korea, the Seed Industry Act was enacted on December 31, 1997, and seed management has been carried out in accordance with the standards of the International Union for the Protection of New Varieties of Plants (UPOV), which Korea joined in 2002 and which includes the plant variety protection system to safeguard plant breeders' rights.

② For all crops, seed quality deteriorates over years of cultivation, making it difficult to maintain the original varietal characteristics. Seeds may also become infected with diseases and pests, reducing productivity. Therefore, seed renewal must be carried out at fixed intervals (for rice, once every four years).

③ As of 2021, the distribution status of certified seeds was as follows: a total of 29 varieties (26 non-glutinous rice varieties and 3 glutinous rice varieties) were supplied, amounting to 15,063 metric tons in total volume. Beginning in 2022, certified seeds have been supplied in an untreated (non-disinfected) form.

④ Seed renewal rates are approximately 60% in Korea, 70% in Japan, and nearly 100% in California, USA.

⑤ The conventional cross-breeding process involves crossing, raising the F_1 (first hybrid generation), segregation and individual selection in the F_2 (second hybrid generation), line development and trait evaluation, yield trials, regional adaptability trials, variety registration and release, seed multiplication, and finally distribution to farmers.

B: Multiplication stages and production institutions in the rice seed supply system

① Breeder seeds (basic plants) are developed and produced by the National Institute of Crop Science (NICS), universities, or private breeding companies.

② Foundation seeds (original seeds) are produced by each Provincial Agricultural Research & Extension Service (ARES) through multiplication of breeder seed. Production costs are partially subsidized by the National Seed Management Office.

Figure 4-3. Rice germplasm field showing rice plants with diverse characteristics (Breeding Field, Kyungpook National University, 2025)

③ Foundation seed is produced at each Provincial Foundation Seed Farm through the multiplication of breeder seed (production costs are subsidized by the National Seed Management Office).

④ Certified seed is produced by multiplying foundation seed from the National Seed Management Office to supply seed for farmers.

C: Seed production technology

① Rice grown for seed purposes is cultivated in fertile paddy fields, using breeder seed and planting one seedling per hill, while removing off-type plants and preventing contamination by other varieties during production.

② Nitrogen fertilizer is applied at a reduced rate to improve grain filling, seed disinfection and pest/disease control are carried out thoroughly, and harvesting is done slightly earlier than normal cultivation (at the yellow-ripe stage).

③ Instead of artificial drying, natural drying is used, and the threshing machine speed is kept below 300 revolutions per minute to minimize seed damage.

2. Rice Breeding and Physiological-Ecological Characteristics

1) Major Characteristics of Rice

(1) Earliness

① Earliness refers to the characteristic of a rice variety in which the maturity period becomes either shorter (earlier) or longer (later).

② It is classified into the following categories: extremely early-maturing, early-maturing, medium–early-maturing, medium–late-maturing, late-maturing, and extremely late-maturing.

③ Earliness is a very important factor in determining the cultivation region and planting season, and it is also one of the breeding objectives.

④ Differences in earliness are primarily determined by the length of the vegetative growth period, because the reproductive growth period shows little variation among varieties.

⑤ The length of the vegetative growth period is mainly influenced by photoperiod sensitivity.

Table 4-7. List of certified rice varieties by maturity group in 2022

Maturity group (Number of varieties)	Variety Name (Main Cultivation Region)
Early-maturing (6)	Koshihikari, Odae (Gangwon), Oryun, Ungwang, Jomyeong1, Haedamssal
Medium-maturing (4)	Matdream, Hiami, Haepum, Boramchal
Medium late-maturing (17)	Daean, Mipum, Samgwang (Chungnam), Saenuri, Saeilmi, Saecheongmu (Jeonnam), Sindongjin (Jeonbuk), Yeonghojinmi (Gyeongnam), Ilmi, Ilpum (Gyeongbuk), Chamdream (Gyeonggi), Chucheong, Chindeul, Hyeonpum, Yeongjin, Dongjinchal, Baegokchal

*Variety names in parentheses indicate the main cultivation region.

A: Climatic–ecological type

① The climatic–ecological type refers to the transition from the vegetative phase to the reproductive phase in response to climatic conditions such as day length and temperature, and therefore it is also called the photothermal development theory.

Table 4-8. Classification of climatic–ecological types

Climatic–ecological type component	Related description	Region where pronounced	Variety characteristics
Basic vegetative growth (B)	• Even if a crop is exposed to suitable temperatures and day lengths for heading and flowering, it will not reach heading and flowering without a certain minimum amount of basic vegetative growth. • The degree of basic vegetative growth is indicated by the minimum number of leaves, and the greater the basic vegetative growth, the higher the minimum leaf number.	Pronounced in low-latitude regions such as Southeast Asia (Blt)	Tongil-type varieties
Photosensitivity (L)	• Short-day plants, in particular, tend to promote heading and flowering under short-day conditions. • The degree of photosensitivity is expressed by the extent to which heading and flowering are promoted under short-day conditions.	Pronounced in southern plain regions of Korea (bLt)	Late-maturing varieties
Thermosensitivity (T)	• The tendency for heading and flowering to be promoted more by high temperatures than by low temperatures, until the optimum temperature for growth is reached. • The degree of thermosensitivity is expressed by the extent to which higher temperatures promote heading and flowering. Thermosensitivity is related to growth temperature and is not directly related to vernalization.	Pronounced in mid-latitude mountainous regions of Korea (blT)	Early-maturing varieties
Combined photosensitivity and thermosensitivity	• A case where day length and temperature have interactive effects on heading and flowering.	In high-latitude regions of Korea, both thermo-sensitivity and photosensitivity are low (blt)	Extremely early-maturing varieties

② Based on the heading and flowering responses of crops to temperature and day length during growth, crop varieties are classified into different varietal groups.

③ In the notation, if an alphabetic character is in uppercase (e.g., B, L, T), the trait is expressed, and if it is in lowercase (e.g., b, l, t), the trait is not expressed.

B: Floral initiation and ecotype

① All rice varieties possess basic vegetative growth, photosensitivity, and thermosensitivity, but the proportion of each trait varies among varieties.

② In early-maturing varieties, thermosensitivity is relatively greater than photosensitivity, whereas in late-maturing varieties, photosensitivity is greater than thermosensitivity.

③ In tropical low-latitude regions such as Southeast Asia, varieties with high basic vegetative growth and very weak photosensitivity are common.

④ In high-latitude or cold regions, early-maturing varieties with short growth durations are suitable, although their yield potential is lower.

C: Application of climatic–ecological types

① If a basic vegetative growth type variety adapted to equatorial regions is cultivated in the southern plain regions of Korea, heading may be delayed, leading to ripening disorders.

② If a late-maturing variety adapted to the southern plain regions of Korea is cultivated in the northern mountainous regions, damage from low temperatures will occur.

③ If a climatic–ecological type variety adapted to the northern mountainous regions of Korea (blt) is cultivated in the southern plain regions, it will respond to high temperatures by heading and flowering earlier than normal.

④ If a climatic–ecological type variety adapted to the southern plain regions of Korea is cultivated in equatorial regions, it will respond to high temperatures and short-day conditions by heading and flowering earlier than normal.

(2) Yield Potential

① Although rice yield is influenced by environmental conditions and cultivation techniques, the inherent genetic characteristics of a variety also have a significant impact. High yield requires strong photosynthetic capacity, slow translocation efficiency to panicles, and low respiratory consumption, even though the heritability of photosynthesis itself is not high.

② Yield and grain quality are negatively correlated, and varieties with high yield tend to have lower grain quality.

③ Ultra-high-yielding varieties (670~740kg/10a) include Anda, Dasan, Namcheon, Areum, and Namil.

④ Another reason for higher yield potential in temperate regions is that the nighttime temperature during the grain-filling period is lower than in tropical regions, reducing carbohydrate loss from respiration. In addition, the grain-filling period is generally longer in temperate regions than in tropical regions.

(3) Plant Type

The plant type of rice refers to the overall combination of traits such as plant height, number of tillers, openness of tiller angles, leaf angle, leaf erectness, number of panicles, and panicle size. Changes in plant type affect light interception efficiency for photosynthesis, planting density, and fertilizer–water management requirements.

Table 4-9. Comparison between panicle-number type and panicle-weight type rice varieties

No.	Category	Tillering type varieties	Heavy panicle type varieties
①	Vigor	Stems are thin, roots are shallow (shallow-root type) with many roots.	Plants are tall with deep-root systems (deep-root type).
②	Tiller	Produce many tillers, resulting in a large number of panicles, but panicles are small and light, and grains are also small.	Panicles and grains are large and heavy, but the number of panicles is small.
③	Cultivation	To achieve high yield, a large number of panicles must be secured, and therefore they are suitable for fertile, warm regions or high-fertilizer-input cultivation, where the effect of tiller ratio is significant.	Because of tall plant height, they are susceptible to lodging, making them unsuitable for fertile soils or high-fertilizer-input cultivation.
④	Suitable region / Application	In poor soils, it is difficult to secure enough panicles, making high yield difficult to achieve.	Suitable for poor soils, late planting, close planting, and double cropping where securing a large number of panicles is difficult.

(4) Quality

① The quality of rice is also determined by its inherent genetic characteristics, with multiple genes involved. Because quality is greatly affected by environmental conditions, the efficiency of breeding for quality is relatively low.

② The genetics of grain quality involve quantitative traits showing continuous variation, controlled by major and minor genes, with inconsistent magnitude and direction of effects, and strong interactions with the environment.

③ The inheritance of white-core and white-belly kernels shows that translucent grains are dominant, but this trait is strongly influenced by environmental conditions and is closely related to grain shape.

④ The inheritance of starch type is such that waxy (wx) endosperm is simply recessive to non-waxy (Wx), while high-amylose is incompletely dominant over low-amylose.

⑤ Protein content exhibits both dominant and additive effects, but its heritability is very low, resulting in low selection efficiency.

⑥ Numerous genes are involved in rice aroma, with large variation depending on variety and environment. Complex inheritance patterns are also affected by phenomena such as xenia (expression of pollen traits in the resulting seed).

(5) Palatability

① High-quality rice varieties with lodging tolerance, disease resistance, and resistance to rice blight can produce high yields under high-fertilizer input, but their eating quality tends to decline.

② Varieties with good palatability are genetically taller, have weaker stems, and are more susceptible to lodging and diseases.

③ Poor palatability results when there is little development of the endosperm's network structure and when a large number of protein granules accumulate in the starch cell walls.

④ A high number of protein granules reduces starch swelling during cooking, lowering stickiness and thus decreasing eating quality.

(6) Plant Height

① Plant height is the length from the ground to the tip of the flag leaf during the vegetative growth stage.

② Culm length is the length of the rice stem from the ground to the panicle node at maturity.

③ The culm consists of nodes and internodes, and their number, length, and thickness vary depending on the variety, fertilization, and cultivation conditions.

④ Culm length and plant height are closely related to lodging, as well as to the amount of leaves and stems, which are the photosynthetic and respiratory organs.

⑤ Panicle length, also called rachis length, is usually around 20cm.

(7) Low Temperature Germinability

① The minimum germination temperature for rice is 8~10°C, but generally 12~13°C is the lower limit for practical purposes.

② Varieties from high-latitude regions, such as in Korea, have higher low-temperature germinability than varieties from low-latitude regions such as the Philippines.

③ Among ecotypes, temperate *japonica* varieties generally have better low-temperature germinability than *indica* varieties.

④ Tongil-type varieties have a low-temperature germination threshold 2~3°C higher than that of temperate *japonica* varieties.

⑤ In high-latitude highland cultivation or in temperate zones for early seedling raising, varieties with high low-temperature germinability are advantageous.

(8) Shattering Habit

① Shattering habit refers to the tendency of grains to fall easily from the panicle at maturity.

② Among varieties, *indica* types have stronger shattering habit than temperate *japonica* types.

③ Tongil-type varieties also have stronger shattering habit than temperate *japonica* types.

④ Within a panicle, grains at the tip are more prone to shattering than those at the base.

⑤ High shattering at harvest can be advantageous for machine harvesting with combines.

⑥ However, if exposed to hail or strong winds at maturity, yield loss is high, and in manual harvesting, significant losses occur during harvesting, drying, and transport.

(9) Premature Sprouting (Pre-harvest Sprouting)

① Premature sprouting refers to the germination of grains while still attached to the panicle at maturity.

② It occurs when rice plants that are more than 25 days past heading are lodged by typhoons or other causes and are exposed to high-temperature and high-humidity conditions.

③ Generally, the risk of sprouting increases when temperatures exceed 20°C and rainfall continues for five days, and submersion due to lodging also promotes sprouting.

④ It is especially severe in poorly drained, poorly ventilated mountain valleys.

⑤ Recently, due to abnormal weather patterns, the frequency of autumn rainfall has increased, leading to more frequent premature sprouting.

⑥ The relationship between pre-harvest sprouting and seed dormancy is that early-maturing varieties tend to have stronger sprouting tendency than late-maturing varieties, and early-maturing varieties tend to sprout more readily than late-maturing varieties.

(10) Direct Seeding Adaptability

① Direct seeding adaptability refers to the suitability of a variety for cultivation in which rice seeds are sown directly in the paddy without transplanting seedlings.

② Such varieties must have good germination and emergence even under deep-water conditions, lodging tolerance, strong low-temperature germinability, and rapid early growth.

(11) Sensitivity to Nursery Period

① Sensitivity to nursery period refers to the tendency of transplanted seedlings, raised for a long time or under high temperatures, to produce vigorous tillering immediately after establishment but head prematurely with only a few tillers. This trait is more pronounced in thermosensitive types and less in basic vegetative growth types (Blt < bLt < blT).

② The phenomenon of premature heading tends to occur when early-maturing varieties are transplanted late. In such cases, abnormal growth results, heading becomes uneven, and both yield and quality decline.

(12) Lodging Tolerance

① Lodging tolerance refers to the ability of a plant to resist stem breakage or bending.

② At maturity, panicles become heavy, and plants are prone to lodging from rain or typhoons, so lodging tolerance is essential for high yields.

③ Lodging is influenced by variety, environment, and cultivation conditions, and generally short-culm varieties are more resistant to lodging.

Figure 4-4. Photograph of severely lodged Nakdong rice (Breeding Field, Kyungpook National University, 2023)

(13) Adaptability for Heavy Fertilizer Application

① Varieties with adaptability for heavy fertilizer application can assimilate nitrogen efficiently even when fertilizer application rates are increased, without growth inhibition or yield reduction.

② In the narrow sense, adaptability refers only to the physiological ability for nitrogen assimilation. In the broader sense, it also includes resistance to diseases and pests and lodging under high-nitrogen conditions.

③ Most high-yielding varieties developed in recent years are adaptable to nitrogen application and show yield increases under such conditions.

④ Such varieties generally have short plant height, erect leaves, and a plant type with good light interception.

(14) Disease Resistance

① Disease resistance refers to the ability of a plant to resist the invasion of pathogens.

② Resistance to invasion is the ability to prevent the entry of a pathogen into the plant, whereas resistance to spread is the ability to restrict pathogen spread within the tissues after invasion.

③ In some cases, pathogens may be present due to latent infection without visible disease symptoms.

④ Tolerance refers to cases in which disease symptoms appear but actual damage is minor.

(15) Insect Resistance

① Insect resistance refers to the ability of a plant to resist the invasion of insect pests.

② Differences among insect-resistant varieties are due to morphological, chemical, and color traits, which influence pest non-preference and tolerance.

③ Loss of resistance refers to the emergence of new susceptible biotypes after resistant varieties have been widely cultivated.

(16) Cold Tolerance

① Cold tolerance refers to the degree of resistance to low-temperature damage (chilling injury) during the growing season.

② Cold damage can occur at any stage of rice growth due to low temperatures.

③ Cold tolerance varies by growth stage, making breeding for this trait challenging.

④ Cold tolerance is important for highland and mountainous areas, cold regions, high-latitude regions, early planting, and late planting.

(17) Drought Tolerance

① Drought tolerance refers to the ability to withstand drought or dry conditions.

② Drought tolerance is expressed through morphological, anatomical, and physiological traits of the rice plant.

③ It is an important trait for varieties cultivated in poorly irrigated paddy fields where drought frequently occurs.

(18) Salt Tolerance

① Salt tolerance refers to resistance to salinity.

② In reclaimed lands or saline soils where salts have accumulated over long periods, varieties with high salt tolerance should be selected.

2) Directions for Variety Improvement

(1) Breeding Objectives

① In Korea, consumer preference is for high-quality rice with stickiness, good chewiness, gloss, and a savory aroma, and therefore breeding should focus on improving eating quality.

② Desirable varieties should be ultra-high-yielding, have improved light interception posture, and enhanced photosynthetic capacity for greater efficiency.

③ To minimize pesticide use, varieties should be developed with broad-spectrum resistance to both diseases and pests (multiple resistance).

④ Varieties should be developed with strong environmental tolerance to withstand multiple stresses from abnormal climate and environmental pollution.

⑤ Functional varieties should be bred for diverse uses.

⑥ The diversity of varieties should be expanded to include rice with health-promoting antioxidant compounds, enriched with specific amino acids or minerals, colored and aromatic rice, extra-long or small-grain rice, and giant-embryo rice.

(2) Considerations in Variety Selection

① Select varieties with high yield and high quality that can express optimal yield and quality under the environmental conditions of each region. Utilize cultivation information from the National Seed Management Office and the Rural Development Administration.

② For direct seeding cultivation, select varieties with strong low-temperature germinability, flooding tolerance during germination, vigorous early growth, deep roots, and short plant height for lodging resistance.

③ For double cropping after barley or wheat, select short-duration varieties or those with strong adaptability to late planting.

④ When selecting functional or specialty rice varieties for processing, ensure market channels in advance by conducting demand surveys and contracting with processing companies for production.

⑤ In areas prone to drought, select varieties with strong adaptability to both direct seeding and late planting.

⑥ When choosing new varieties, carefully review adaptability to the region, recommended fertilization rates, and resistance to diseases and pests.

⑦ In suburban, roadside, or factory-adjacent areas with nighttime lighting, avoid medium–late maturing varieties and select early-maturing varieties to prevent heading delays.

⑧ In each region, cultivate 2~3 recommended varieties with different heading dates in a staggered schedule to improve farm machinery efficiency, ensure timely harvest, and spread the risk of various disasters.

Rice Crop Science

Chapter 5. **Cultivation Environment of Rice**

Chapter 5.
Cultivation Environment of Rice

Although the previous chapter discussed rice varieties, the natural environment of a given region or country ultimately plays a decisive role. Overcoming environmental constraints is possible to some extent, but for sustainability and efficiency, breeding must also focus on improving environmental adaptability. While cultivation techniques can also help overcome environmental limitations, they must likewise conform to natural conditions.

Korea is traditionally a rice-cultivating country. The abundance of water resources and the hot, humid climate have been suitable for making rice the main crop. Although many crops can be grown, few match Korea's cultivation environment as well as rice, barley, and soybeans. Being complementary to each other and well-suited to the climate, they have enabled the development of diverse farming techniques over a long period. In this chapter, we examine the climatic and soil environments.

1. Climatic Environment

1) Climatic Environment and Rice Cultivation

(1) Temperature

① Korea is located in relatively high latitudes and has many mountainous areas, and low temperatures are a major constraint to stable rice cultivation.

② Rice growth is the result of physiological and biochemical processes in which enzymes play a key role, and their activity is strongly influenced by temperature.

③ Rice growth varies with variety, temperature, light intensity, plant age, and nutritional status.

④ The minimum temperature for rice growth is about 10 °C, the optimum range is 25~35°C, and the maximum is around 40°C, and temperatures above or below these thresholds impair growth.

⑤ Cumulative temperature (growing degree days) is the sum of daily mean temperatures over the growing period. Although it varies by region and variety, rice generally requires about 2,500~4,500°C.

⑥ As long as the daily minimum temperature does not fall below the critical threshold, a larger diurnal temperature range is favorable for tillering and grain filling.

⑦ During the tillering stage, the growing point is still underwater, so early growth is more influenced by water temperature than air temperature.

⑧ From panicle initiation to booting stage, the growing point rises to near the water surface due to internode elongation, so air temperature and water temperature have similar effects on yield.

⑨ During grain filling, most parts of the plant are above the water surface, making air temperature more influential than water temperature.

(2) Light Intensity

① Strong solar radiation allows light to penetrate into the inner canopy, enhancing photosynthesis and dry matter accumulation, thereby increasing yield.

② If excessive nitrogen fertilizer is applied after maximum tillering, the canopy becomes too dense, reducing canopy photosynthesis and lowering yield.

③ Canopy photosynthesis is more strongly affected by the interaction of light and temperature than by light intensity alone.

④ Within the optimal temperature range for growth, both higher temperatures and greater light intensity increase photosynthesis.

⑤ If temperatures exceed the optimum, higher light intensity can actually reduce photosynthesis through increased photorespiration.

⑥ Day length affects photosensitivity and thus heading, and rice heads earlier under short-day conditions.

(3) Water

① Rice requires sufficient water for good establishment after transplanting, vigorous growth, improved resistance to pests and diseases, and reduced weed occurrence.

② Adequate water supply facilitates puddling, plowing, leveling, and transplanting operations.

③ In the era of rainfed paddies, spring rainfall was the greatest limiting factor, but with the development of modern irrigation facilities, if irrigation water is sufficiently supplied, the absence of rainfall can actually be advantageous because it increases solar radiation and air temperature while lowering humidity, thereby promoting transpiration and increasing yield.

④ Excessive and frequent rainfall reduces photosynthesis, weakens plants, increases susceptibility to pests and diseases, promotes lodging, and raises the risk of pre-harvest sprouting during grain filling, lowering yield.

(4) Air

① From maximum tillering to early grain filling, excessive canopy density reduces ventilation, limiting photosynthesis due to CO_2 deficiency.

② Because rice absorbs large amounts of CO_2, temporary CO_2 depletion within the canopy is common during vigorous growth.

③ Light wind supplies fresh air with CO_2 into the canopy, promoting photosynthesis through the border effect.

④ Light wind also lowers humidity, enhancing transpiration, nutrient uptake, and reducing pest and disease incidence, thus benefiting growth.

⑤ Strong winds stress plants, induce stomatal closure, and cause mechanical injury that facilitates pathogen entry.

⑥ Excessively strong winds can disrupt flowering and pollination, and in severe cases, cause lodging damage.

⑦ High-temperature, dry winds or strong winds before and after heading can cause white head symptoms, where panicles dry out and die.

2. Nitrogen Cycle in Paddy Soils

1) Transformation of Organic Nitrogen in Paddy Soils

① Ammonification: In paddy soils, organic nitrogen is converted into ammonia through the process of ammonification.

② Nitrification: Ammonium nitrogen is oxidized in the oxidized layer at the soil surface, transforming into nitrate nitrogen.

③ Because nitrate nitrogen is an anion, it is not adsorbed onto soil colloids and is therefore leached downward into the soil profile.

④ Denitrification: The leached nitrate nitrogen, upon reaching the reduced subsoil layer, is converted into nitrogen gas through the action of denitrifying bacteria, and is then lost to the atmosphere.

Table 5-1. Characteristics of flooded paddy fields

No.	Item	Description
①	No continuous cropping injury	In paddy fields, rice can be cultivated continuously every year without a decline in productivity. This is because flooding prevents the survival of pathogens and pests that cause continuous cropping injury, harmful substances are washed away with irrigation water, and a considerable amount of nutrients is supplied from the irrigation water itself.
②	Soil reduction and methane gas emission	During rice cultivation, fields are flooded, and even in the non-cropping winter season, the soil remains wet with limited oxygen supply, leading to reduced (anaerobic) soil conditions. Under such conditions, organic matter decomposition cannot proceed oxidatively, resulting in the emission of methane gas (CH_4), which contributes to global warming and partially to ozone layer depletion.
③	Prevention of salt accumulation and conservation of the natural ecosystem	Flooded paddies function as a fertility-preserving system for rice cultivation. Continuous rice cultivation in flooded fields prevents the build-up of salts, thus avoiding salt accumulation. They also provide habitats for wild freshwater fish, insects, and birds, maintain natural landscapes, and offer benefits for public health and recreation.

2) Nitrogen Fertilization in Paddy Soils

① In paddy soils, deep placement of fertilizer is necessary to prevent denitrification, but because this is difficult to implement in practice, mixed-layer application is used instead.

② Although deep placement is ideal, it is practically difficult, and therefore after applying nitrogen fertilizer, deep plowing is carried out to ensure the fertilizer is evenly mixed into both the oxidized layer and the reduced layer of the soil, and this method is called through-layer application (mixed-layer application).

③ Even with through-layer or mixed-layer application, nitrogen loss is minimal in paddy soils because the oxidized layer (topsoil) is thin and the reduced layer (subsoil) is thick.

④ In paddy soils, ammonium-based fertilizers are preferable to nitrate-based fertilizers, as they are more effective in reducing denitrification losses.

⑤ If ammonium nitrogen is placed directly into the reduced layer from the start, it is well adsorbed by the soil, increasing nitrogen use efficiency.

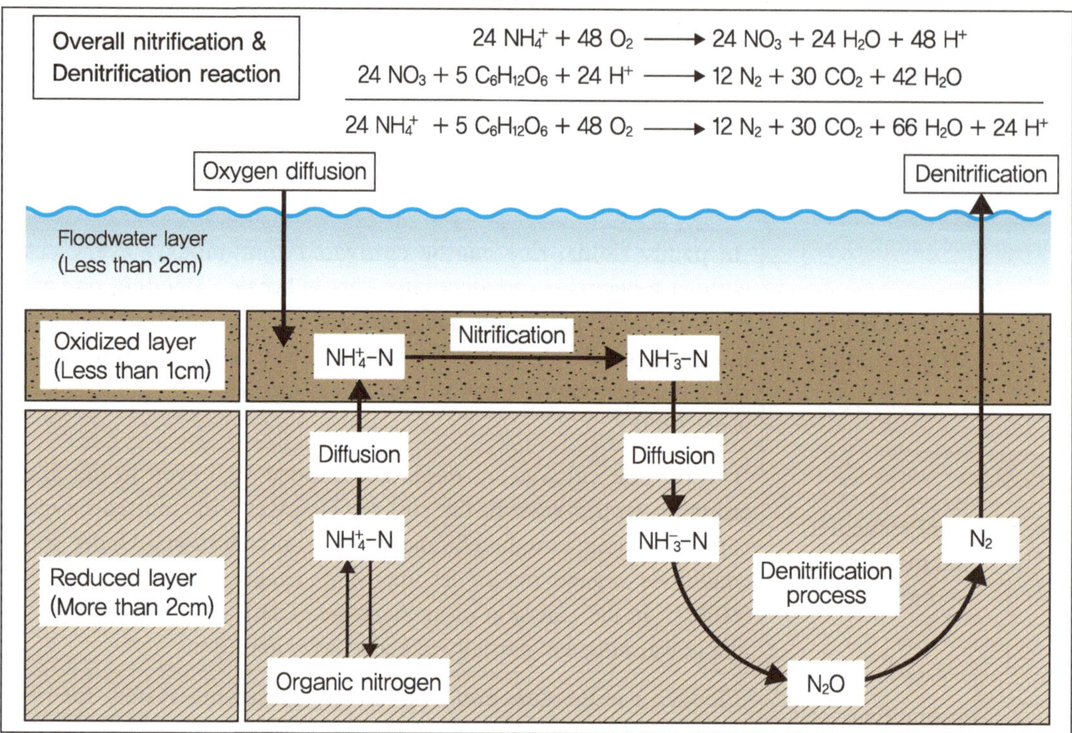

Figure 5-1. Schematic diagram of nitrification and denitrification

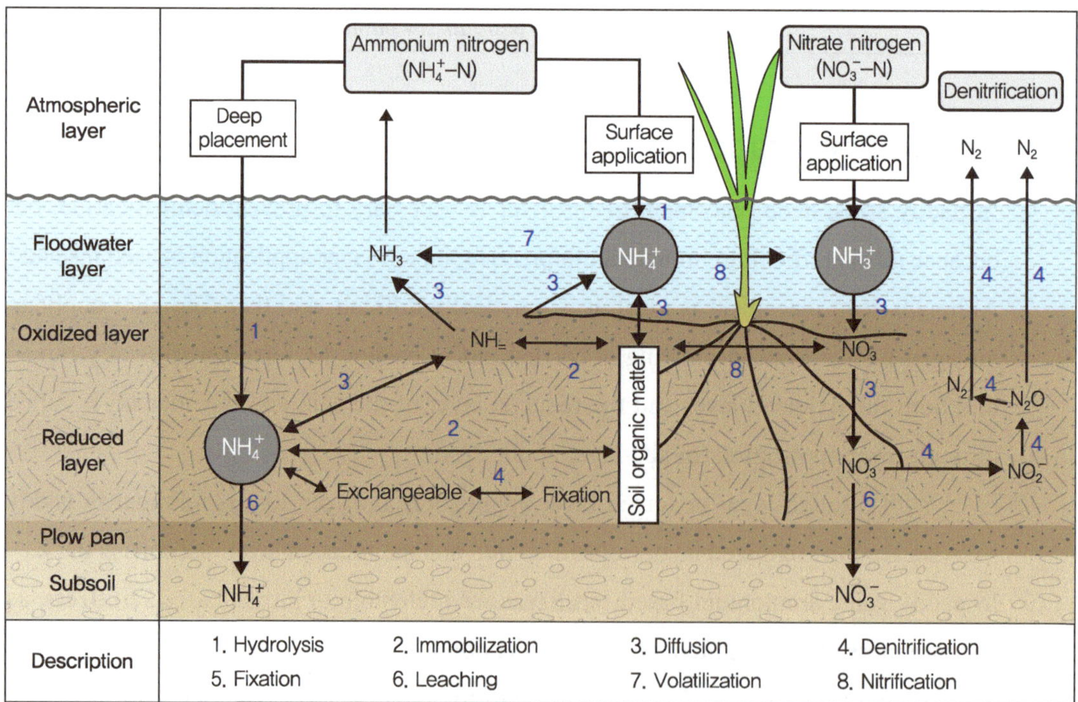

Figure 5-2. Nitrogen cycling and transformation in paddy soils

3. Types and Characteristics of Paddy Soils

1) Types of Paddy Soils

(1) Dry Paddy Field

① A dry paddy field is land that can be used as paddy under irrigation and as upland under drainage.

② About one-third of Korea's paddy fields are dry paddies, also called ordinary paddy fields, and rice produced here has the best eating quality.

③ Dry paddy fields are highly productive and allow for rice–barley double cropping, rice–upland crop rotation, and other crop rotations.

④ Because of the rapid decomposition of organic matter, potential soil fertility is easily reduced, and therefore large amounts of organic matter should be applied, soil conditioners rich in mineral nutrients should be incorporated, and deep plowing should be conducted to increase yield.

(2) Sandy Paddy Field

① A sandy paddy field is characterized by high sand content and rapid water percolation.

② Due to severe seepage, nutrient retention is poor and large amounts of irrigation water are required.

③ Soil improvement involves soil dressing with clay-rich soil.

④ Fertilizers should be split-applied, and slow-release fertilizers are effective in improving fertilizer use efficiency.

(3)Immature Paddy Field

① Recently reclaimed paddy fields with low organic matter content and poorly drained clay loam or clay soils.

② Soil texture is dense, nutrient content is low, and percolation is poor.

③ Improvement involves applying well-decomposed compost in large amounts and deep plowing to improve soil physical properties.

(4) Poorly Drained Paddy Field (High-Moisture Paddy)

① Poor drainage maintains water-saturated conditions, also called high-level paddies.

② Excess soil moisture lowers soil temperature, creates reducing conditions, and gives the soil a dark bluish color.

③ Low temperature slows organic matter decomposition, leading to humus accumulation and poor physico-chemical properties.

④ Rice root development is poor, and in summer high temperatures accelerate organic matter decomposition, aggravating soil reduction, producing toxic gases such as hydrogen sulfide (H_2S), which impair root function and cause autumn decline.

⑤ Improvement involves subsurface drainage to lower the groundwater level and avoiding or reducing application of undecomposed organic matter.

(5) Autumn-Declining Paddy Field (Deteriorated Paddy)

① Paddy soils that have undergone long-term degradation, resulting in yield decline in autumn.

② Autumn decline occurs when iron (Fe) is deficient and rice roots appear whitish due to H_2S toxicity, and plants grow normally during vegetative growth but, during reproductive stages, especially after heading, lower leaves senesce prematurely, root vigor decreases, and 'kesse-mottle' disease occurs, leading to reduced yield.

③ It occurs mainly in old paddies, sandy paddies with severe seepage and low nutrient retention, gravelly paddies (stony paddy fields), and poorly drained paddies with excessive organic matter accumulation.

④ Under flooded conditions, sulfate is reduced to H_2S in the reduced layer, and when Fe is abundant, rice roots form a thick ferric oxide coating that protects against H_2S toxicity by precipitating it as FeS.

⑤ Countermeasures include soil dressing with nutrient-rich soil, application of soil amendments and organic matter, and deep plowing. Excessive or premature application of soluble fertilizers under shallow topsoil conditions aggravates the problem.

(6) Saline Paddy Field

① Paddy fields in reclaimed land with high salinity that restrict normal rice growth.

② Fine-textured soils with poor aeration, strong reducing conditions, and accumulation of toxic gases also experience severe nutrient leaching during desalinization.

③ At salinity levels below 0.1%, rice cultivation is not significantly affected, at 0.1~0.3% cultivation is possible but there is a risk of salt injury, and at levels above 0.3% normal rice cultivation becomes difficult due to salt damage.

④ The best improvement is frequent irrigation for desalination, and the application of straw or lime also promotes desalination.

⑤ Fertilization: ammonium sulfate (N), superphosphate (P), and potassium sulfate (K) are effective.

(7) Soils with Specific Element Deficiencies or Excesses

① Due to parent material characteristics, some Korean paddy soils are inherently deficient in specific elements, while others are affected by contamination from mining or industry.

② Zinc (Zn) deficiency often occurs in limestone areas or saline soils, and improvement requires the application of zinc sulfate at 3kg per 10 ares.

③ Heavy metal–polluted paddies occur near mines or industrial wastewater, with damage from Cd, Cu, As, Pb, Zn, etc.

④ Improvement involves soil dressing with uncontaminated soils, or application of lime and organic matter to prevent heavy metal uptake by rice.

4. Improvement of Paddy Soils

1) Deep Plowing

① Deep plowing refers to tillage in which the plow layer is worked to a depth of 18cm or more.

② Applied to ordinary or immature paddies where shallow topsoil causes fertilizer effect to diminish early or autumn decline to occur.

③ Deep plowing combined with organic matter application and 20~30% increased fertilizer use leads to higher yields.

④ It has little effect on sandy paddies and may be unfavorable in sandy, leaky, or poorly drained paddies.

2) Soil Dressing

① Incorporating soil of different texture to improve paddy soils.

② For excessively drained sandy soils, clay content should be increased to ≥15%.

③ For soil dressing to improve sandy soils, it is desirable to use soil with a clay content of 25% or more, but if this is unavailable, soil with a clay content of at least 20% is also acceptable

④ For poorly drained heavy clay, sand is applied to reduce clay content to ~15%.

⑤ After soil dressing, deep plowing (>18cm) plus organic matter application and slightly increased fertilization are necessary for yield increase.

3) Application of Organic Matter

① Improves soil physical properties, supplies nutrients during decomposition, and enhances enzyme activity.

② Increases fertilizer use efficiency by adsorbing nutrients and pesticides, reducing herbicide injury.

③ Darker color increases soil temperature and improves rice quality.

④ In paddy fields with an organic matter content below 2.5%, rice straw or compost should be finely chopped, incorporated into the soil, and followed by deep plowing.

⑤ Since compost or farmyard manure is difficult to prepare, it is desirable to return rice straw at harvest.

⑥ Excess organic matter may lead to excess nitrogen in late growth stages, causing lodging and poor eating quality.

⑦ Most Korean paddy soils cannot exceed 3% organic matter content even after long-term application.

⑧ To reduce lodging risk, basal fertilizer ratio should be lowered and topdressing ratio increased when applying organic matter.

⑨ Livestock manure can also be applied fresh, but should be incorporated immediately after application in April.

4) Green Manure Cropping

① Essential for high-quality rice production by enhancing soil fertility.

② Green manure crops such as rye, hairy vetch, and Chinese milk vetch are cultivated.

③ After flooding in transplanting season, rotary tillage accelerates decomposition before transplanting.

④ Application of silicate or lime fertilizers promotes decomposition of green manure.

⑤ For gramineous green manure crops, additional N (3~4 kg/10a) is recommended.

5) Application of Silicate Fertilizer

① Silicate fertilizers strengthen culms and leaves of gramineous crops, reducing lodging.

② In rice, Si improves light interception and photosynthesis, enhancing yield and quality.

③ Silicate application increases ripening ratio and grain filling, improving yield and grain quality.

④ Since many paddy fields in Korea are deficient in silicon due to insufficient natural supply, silicate fertilizer should be applied to fields with a silicon content below 130ppm.

⑤ Applied to fields with recurrent pests, diseases, cold injury, or lodging, and every 4 years after prior application.

⑥ Application timing: before autumn or spring plowing, at least 2 weeks before basal fertilization.

5. Nutritional Conditions and Rice Quality

① For high-quality rice, organic fertilizers must be applied to improve soil fertility.

② Poorly drained paddies are more fertile than dry paddies, but nitrogen uptake is delayed until late stages, lowering yield and eating quality.

③ Regional differences in rice quality are largely due to soil type, cultivation season, and fertilization practices.

④ Reduced dependence on N fertilizer promotes balanced growth, enhancing flavor and palatability.

⑤ For good taste and aroma, rice must absorb micronutrients evenly in addition to macronutrients.

⑥ Application of organic fertilizers supplies these micronutrients, and if they are unavailable, straw incorporation is beneficial.

⑦ Soil dressing or silicate fertilizer application also favors high-quality rice production.

Rice Crop Science

Chapter 6. **Rice Cultivation Technology**

Chapter 6.
Rice Cultivation Technology

Even with good varieties and favorable climatic conditions, diverse cultivation techniques are required. No matter how excellent the variety or how suitable the environment, without properly combining inputs, cultivation methods, and seasonal timing, only empty grains may result. This highlights the importance of performing cultivation according to the appropriate season neither too early nor too late. The 24 Solar Terms serve as an excellent agricultural calendar. While the lunar calendar was traditionally used, this solar-based agricultural calendar was exquisitely designed to align with solar energy utilization. The ultimate product of rice cultivation is carbohydrates, synthesized through photosynthesis from carbon dioxide (i.e., fertilizer) and water using solar energy. At this stage, nutrient supply and water management are crucial, especially because rice is a crop highly dependent on water as an environmental resource. In addition, traditional cultivation techniques must now be integrated with agricultural mechanization. The mechanization rate for rice has reached 98%, reflecting the fact that mechanization now largely determines cultivation preferences. While reducing fertilizer and pesticide use is necessary, they remain essential inputs. Going forward, detailed and integrated techniques must be applied to continuously improve rice cultivation practices.

1. Rice Cultivation Systems

1) Cultivation Systems

(1) Transplanting Cultivation

① Since the Joseon Dynasty, this method has involved raising seedlings in nurseries (seedbeds) and then transplanting them into paddy fields.

② Hand transplanting refers to transplanting rice seedlings manually, which was the predominant practice until the 1970s.

③ Machine transplanting began in 1977 with the introduction of transplanting machines, and today nearly all transplanting is mechanized.

(2) Direct Seeding Cultivation

① A cultivation method in which rice seeds are directly sown in paddy fields or upland fields.

② Direct seeding on dry paddy involves sowing on dry land and then flooding after seedling emergence.

③ Direct seeding on flooded paddy involves sowing seeds into already flooded fields.

④ When direct seeding was first introduced, dry direct seeding was preferred because of operational convenience, but since the mid-1990s, direct seeding on flooded paddies has been increasing due to abnormal rainfall during the spring sowing season. In Japan as well, direct seeding, which had been in decline, has been increasing again for the cultivation of small-scale functional rice varieties.

2) Transition of Cultivation Practices

① From the Joseon Dynasty until the 1970s, most rice cultivation relied on hand transplanting.

② In the 1980s, machine transplanting of young seedlings was adopted to reduce labor requirements, and direct seeding also began.

③ By the 2000s, hand transplanting had nearly disappeared, with mid-seedling machine transplanting accounting for 76%, young-seedling machine transplanting 15%, direct seeding 8%, and upland rice cultivation 1%.

④ By the 2010s, hand transplanting was virtually absent, direct seeding was rarely practiced, and machine transplanting had become the dominant method.

3) Seed Preparation

(1) Sowing Process (Pretreatment)

The steps are variety selection, seed selection, seed disinfection, seed soaking, pregermination, sowing, germination, greening, and hardening.

(2) Variety Selection

① The variety should be a superior one suited to the climatic characteristics and intended use of the cultivation region, preferably produced in the previous year, and with a germination rate of at least 98%.

② Disease-free seeds should be obtained from certified seed sources such as the Korea Seed & Variety Service or Agricultural Technology Centers.

③ Although rice is a self-pollinated crop, its natural outcrossing rate is less than 1%, and occasional mutations or inadvertent seed mixtures may occur, so seeds are renewed every four years.

④ To prevent damage from pests, diseases, or environmental stresses, resistant varieties should be chosen.

(3) Seed Selection

① Until rice seedlings grow to the 3~4 leaf stage, they depend on nutrients stored in the endosperm, and to obtain vigorous seedlings, well-matured and heavy seeds must be selected.

② Salt water selection is a method of separating well-filled (heavier) seeds by floating them in saline solution.

③ For salt water selection, the specific gravity of the solution is adjusted to about 1.13 for non-glutinous rice, 1.10 for bearded rice, and 1.08 for glutinous and upland rice.

④ Traditionally, the density was estimated by dissolving salt or ammonium sulfate in water until an egg floats, then using the solution for seed selection.

⑤ Today, hydrometers are commonly used for more convenient and accurate seed selection.

(4) Seed Disinfection

① Seed disinfection is conducted to prevent seed-borne diseases such as bakanae, blast, seedling blight, and bacterial grain rot.

② It can be performed either before or after soaking, but pre-soaking disinfection is preferred to reduce the risk of phytotoxicity during germination.

③ Effective solution temperatures range from 10~30°C, with 30°C recommended for bakanae disease control.

④ In regions with stem nematode problems (Aphelenchoides besseyi, causing white tip disease), nematicide treatment should be combined.

⑤ Since 2022, certified seed supplied by the government has been distributed without pre-disinfection, leaving farmers responsible for treatment.

(5) Seed Soaking

① Seed soaking refers to immersing rice seeds in water to absorb the moisture necessary for germination.

② It is conducted after seed disinfection to ensure water uptake and remove germination inhibitors present in the seed coat.

③ Seed soaking shortens the period to germination, reduces germination disorders, promotes uniform seedling establishment, and accelerates early growth.

④ For wet-bed nurseries, soaking prevents seeds from floating or shifting.

⑤ The minimum soaking period required for germination is about one day, though the rate of water absorption increases with water temperature.

⑥ The cumulative soaking temperature for seed soaking is set at 100°C, and this index is used by farmers for large-scale soaking and pregermination.

⑦ At a water temperature of 10°C, soaking requires about 10 days, at 15°C it requires 5 days, at 20°C it takes 3~4 days, and at 25°C about 3 days. When soaked in 30°C water for 2 days, 95% of the seeds germinate, but excessively high temperatures reduce germination quality.

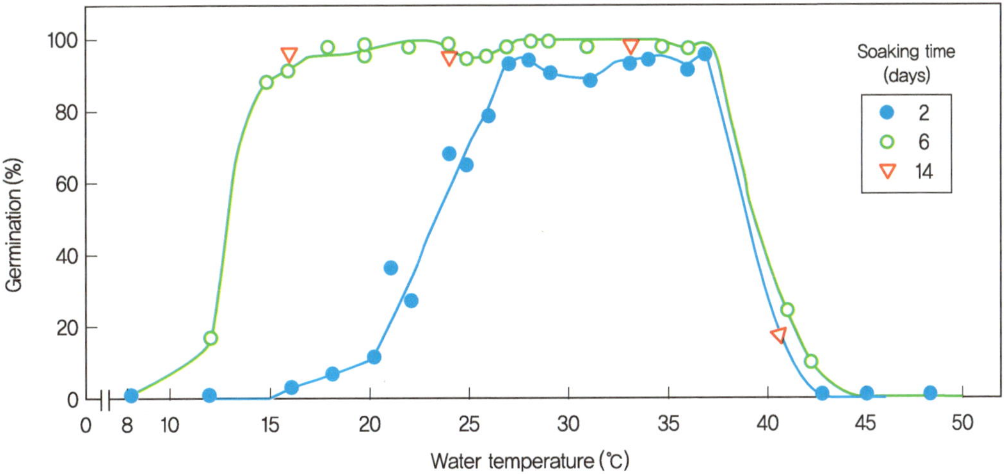

Figure 6-1. Changes in germination rate as affected by soaking duration and water temperature (Livingston and Haasis, 1933).

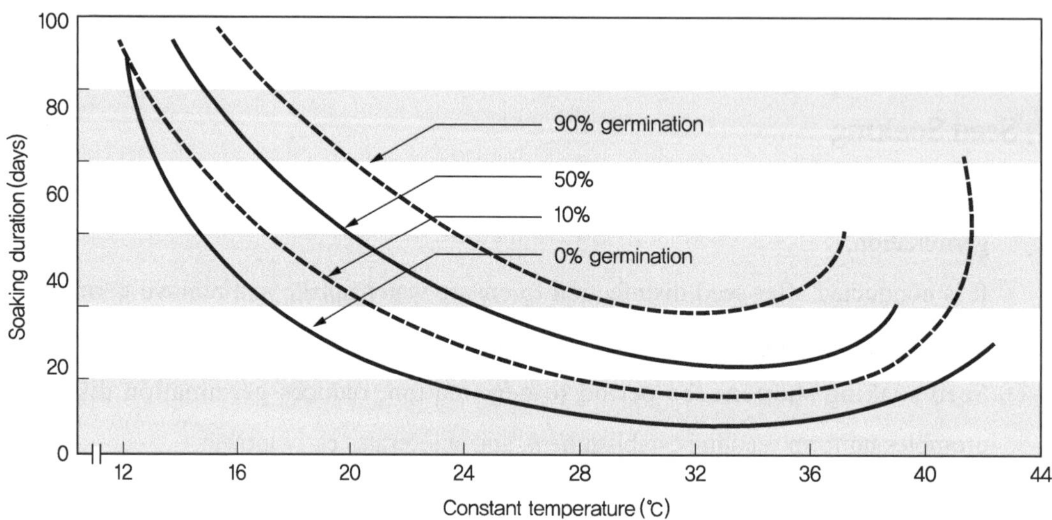

Figure 6-2. Relationship between soaking duration and germination percentage under constant temperature conditions (Livingston and Haasis, 1933).

(6) Pregermination (Hastening of Germination)

① Pregermination refers to sprouting seeds under optimal conditions after soaking so that the absorbed water ensures rapid and uniform germination.

② After soaking, seeds are removed from the water, wrapped in a wet cloth to maintain saturation, and placed at 30°C, and within about 2 days, sprouts emerge. For Tongil-type varieties, pregermination usually takes 1~2 days longer than for *japonica* varieties.

③ Pregermination should continue until the coleoptile reaches about 1 mm, since longer growth makes it fragile and difficult to handle.

④ In practice, seed disinfection and soaking are done 4~5 days before sowing, and pregermination is initiated 2~3 days before sowing.

4) Seedling Raising

(1) Seedling Raising Methods and Characteristics

A: Seedling raising in seedbeds

① Seedbeds are plots for raising seedlings. Seedling raising in field seedbeds is the traditional method for hand transplanting, while tray seedling raising is designed for machine transplanting.

② Water seedbeds are managed under flooded conditions, upland seedbeds are managed under non-flooded upland conditions, alternate seedbeds combine both conditions depending on management needs, and protected alternate seedbeds are covered with plastic film or other materials for heat retention.

③ Generally, seedbed raising is practiced outdoors, whereas tray seedling raising is carried out indoors in multi-layer stacked systems.

④ For efficiency and uniformity, seedlings are often raised cooperatively and purchased as communal nursery stock.

Table 6-1. Characteristics of seedlings by type

No.	Item	Machine-transplanted seedlings		Hand-transplanted seedlings	
		Young seedlings	Semi-young seedlings	Middle-aged seedlings	Matured seedlings
①	Consumption of endosperm nutrients (%)	50	80	100	100
②	Seedling age (leaf number)	2	3.3	4.5~4.8	6.5
③	Node of tiller emergence	2	2	5	5
④	Number of tillers (per hill)	36	32	27	10
⑤	Number of nursery trays (per 10a)	15	20	30	not applicable (N/A)
⑥	Seedling days in nursery (days)	10	20	30~40	40~45
⑦	Establishment ability at low temperature	High	High	Moderate	Low
⑧	Plant height (cm)	7~10	12~14	17~18	22~24
⑨	Delay in heading (days)	4	2	0	0
⑩	Seeding rate (g/tray)	200	150	100	300

B: Tray seedling raising

① This method involves filling small plastic trays with nursery soil, sowing seeds densely, and forming a root mat suitable for machine transplanting.

② Unlike field seedbed seedlings, tray seedlings can be produced in any tray material, as long as they meet the density and matting requirements of transplanting machines.

③ Sowing 200 g of seed per tray produces about 7,000 seedlings, while 220g produces about 8,000 seedlings.

(3) Advantages and Disadvantages of Young Seedlings for Machine Transplanting

A: Advantages

① Since 30~50% of endosperm nutrients remain in the seed, transplanting shock is reduced and root establishment is faster.

② They have high cold tolerance, strong adaptability to environmental conditions, and high submergence tolerance, allowing good growth even under flooded conditions.

③ Because seedlings are small and transplanted shallowly, tillering is promoted.

④ The seedling-raising period is shortened, reducing labor input for nursery management.

⑤ Nursery area and input requirements are reduced.

⑥ The use of transplanting machines reduces labor at the time of transplanting.

B: Disadvantages

① Heading is delayed by 3~5 days compared with medium seedlings, and therefore transplanting must be done earlier.

② The suitable transplanting window is narrow, so timing must be managed carefully.

③ Because seedlings are short, the paddy surface must be well-leveled.

④ Young seedlings have low herbicide tolerance, requiring careful application.

C: Variety selection for machine-transplanted young seedlings

① Varieties with strong submergence tolerance and lodging resistance should be selected.

② In rice–barley double cropping, late-planting adaptable varieties are preferable to early-planting types.

③ Early-maturing varieties are preferred over late-maturing ones.

(4) Advantages and Disadvantages of Water-Seedbed Seedling Raising

A: Advantages

① At low temperatures, the water layer provides a warming effect.

② With intensive management, seed germination and growth are uniform.

③ Weed incidence and damage from rodents, birds, and pests are reduced.

B: Disadvantages

① Compared with upland seedbeds, oxygen deficiency restricts root growth.

② Seedlings tend to become weaker overall, causing greater transplanting shock and lower adaptability to late planting.

(5) Advantages and Disadvantages of Upland Seedbed Seedling Raising

A: Advantages

① Compared with water seedbeds, seedlings are shorter but sturdier, with higher nitrogen and starch content, resulting in stronger root development and greater drought tolerance.

② After transplanting, seedlings suffer less transplanting shock and exhibit rapid early growth.

③ Favorable under conditions where rice growth tends to be delayed, such as in cool regions or fertile soils.

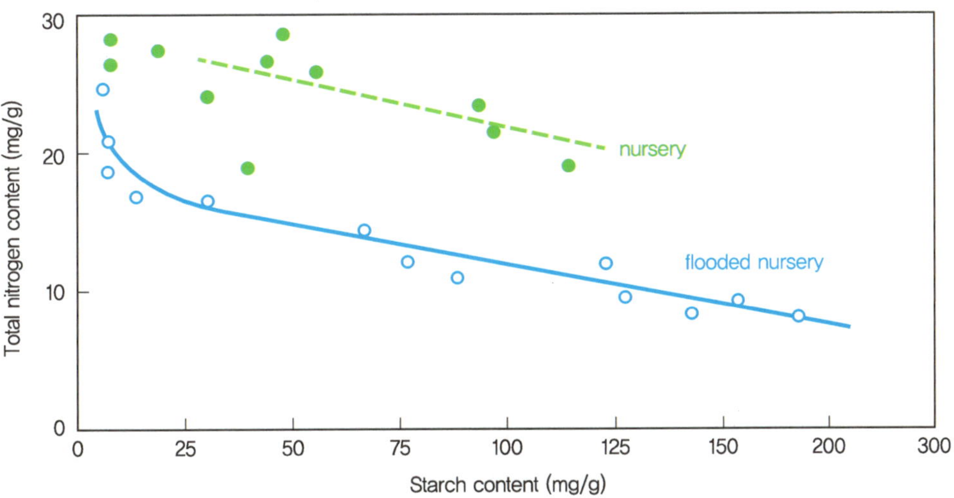

Figure 6-3. Correlation between total nitrogen content and starch content (Yamada and Ota, 1957).

B: Disadvantages

① Upland seedlings absorb less silicon, leading to weaker cell silicification, elongated plant type, and higher susceptibility to rice blast disease.

② Weed incidence is high, and seedlings are more vulnerable to damage from rodents and birds.

③ Germination and growth are less uniform compared with water seedbeds.

(6) Alternate Seedbed Seedling Raising

① This refers to a seedbed method that combines the advantages of both water seedbeds and upland seedbeds.

② In some cases, seedlings are raised under water-seedbed conditions in the early nursery stage and then switched to upland-seedbed conditions in the later stage, or vice versa.

(7) Protected Alternate Seedbed Seedling Raising

① This is an alternate seedbed system covered with plastic film for heat retention.

② It was rapidly disseminated beginning in 1972, when Tongil-type rice varieties (which are sensitive to low temperatures) were introduced, and by 1978 it accounted for 90% of all nurseries. Even today, it remains the most widely used seedbed system.

③ It is particularly effective in cold regions for early planting in double cropping systems or for high-yielding cultivation.

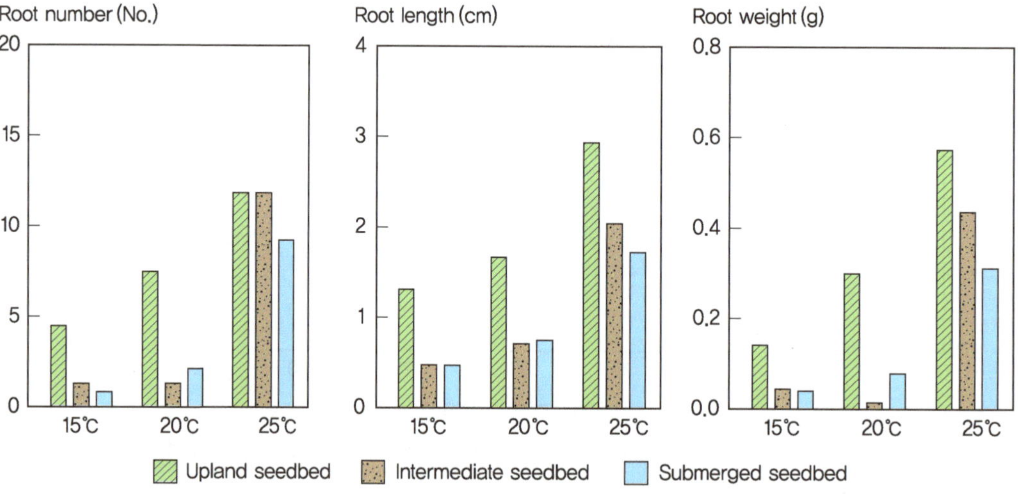

Figure 6-4. Differences in seedling growth according to temperature and type of seedbed (Yatsuyanagi, 1960).

(8) Non-woven Fabric Seedbed Nursing

① In mechanical transplanting of medium seedlings, seedlings are covered only with non-woven fabric without vinyl film for simple insulation.

② Suitable non-woven fabric is about 30–40 g/m² with approximately 30% shading capacity.

③ An advantage is that, compared with using plastic film or wooden laths for vinyl tunnels, this method reduces the labor required for seedling raising and can prevent high-temperature injury.

④ A disadvantage is that, because the insulation effect is weak, it is not suitable for early planting or for regions with heavy rainfall, and furthermore, since it is installed flat over the nursery boxes, heavy rain may cause submergence and lead to poor germination.

⑤ It is safer to apply this method in southern plains than in northern regions.

(9) Dry Seedbed Nursing

① For mechanical transplanting, the field is plowed in a dry condition, and nursery boxes are set directly on the soil. After sowing and soil covering, the seedbed is covered with vinyl film, followed by irrigation.

② Advantage: Field operations are easier since the work is carried out on dry paddy fields.

③ Advantage: Because nursery boxes are transported mechanically rather than manually, labor required for irrigation, pregermination, and box setting is greatly reduced.

④ Disadvantage: Seedling emergence may become uneven, and the rate of healthy seedling establishment tends to be low.

2) Nursery Methods

(1) Box Seedling Nursing

① Nursery boxes of 60 × 30 × 3cm are used, in accordance with transplanter specifications.

② The shallow depth allows roots to grow in a mat-like structure with minimal soil.

③ Small holes at the bottom allow rice roots to absorb water and nutrients from the soil. If too large, separation at transplanting becomes difficult, and if too small or few, water absorption and drainage become poor.

(2) Bed Soil

① Bed soil may be self-prepared or purchased commercially.

② It should be clayey, well-drained, and suitable for root development, with proper humus content, water retention, and free from pathogens, pests, or weed seeds.

③ The optimum soil pH is about 5.0, and should remain within 4.5~5.5. Higher pH makes damping-off more likely, and in such cases, sulfur powder may be applied to adjust pH and suppress pathogens.

④ Fungicides for damping-off and patchy seedling disorder should be applied (e.g., Dazomet, Validamycin, Allclean).

⑤ Damping-off frequently occurs within 2 weeks after germination, mainly due to low temperature.

⑥ Patchy seedlings occur when seedlings lose water balance due to sudden changes in nursery bed temperature.

⑦ This disorder often appears at the 3-leaf stage when seedlings are repeatedly exposed to low temperatures or after sudden warming following a cold period.

⑧ Control measures include promoting root development and preventing rapid fluctuations in soil moisture and temperature.

⑨ In rice, patchy seedling disorder and damping-off are likely to occur when the soil pH rises above 6 and approaches alkalinity, and therefore caution should be taken in applying lime or alkaline fertilizers.

3) Sowing Methods

(1) Seed Disinfection

① Use disease-free seeds, and conduct thorough seed disinfection and salt-water selection.

② Pay attention to seed-borne fungal diseases such as rice blast, brown spot, bakanae disease, as well as bacterial seedling blight.

③ When damping-off or patchy seedlings occur, apply appropriate chemicals and prevent the occurrence of albino seedlings.

④ Floating seedlings occur easily when over-irrigated, excessively dried, or covered with very fine soil, and therefore water management must be carefully controlled.

⑤ Dense sowing in nursery boxes creates favorable conditions for disease development, and once disease occurs, the spread is rapid.

(2) Sowing

① The appropriate sowing time is 25~30 days before transplanting.

② The general sowing rate is 100~130g (about 4,000 seeds) per box, depending on the 1,000-grain weight.

③ For broadcast sowing: small-grain varieties (20~22g/1,000 grains) are sown at 100~120 g/box, medium-grain varieties (23~25g) at 120~130g/box, and large-grain varieties (26~28) at 140~150g/box. For drill sowing, the sowing rate is reduced.

④ For young seedlings, dense sowing of about 220g per box (≈8,000 seeds) is recommended to promote root mat formation.

⑤ Recently, multi-tier shelf systems in indoor facilities are increasingly used, which allow environmental control, produce high-quality seedlings, and save labor.

4) Seedling Management

(1) Seedbed Nursing

① Conventional nursery bed dimensions (submerged seedbed, upland seedbed, intermediate seedbed, vinyl-covered intermediate seedbed) are 120cm in width with 30~40cm paths, and for hand transplanting, vinyl-covered intermediate seedbeds are most commonly used.

② The sowing rate is about 80g/m² (≈ 3,600 seeds).

③ Fertilization: apply 40g N, 30 g P_2O_5, and 50g K_2O per 3.3m².

④ For 10 a of paddy field, about 40m² (12 pyeong) of sowing area and about 66m² (20 pyeong) including paths are required, which is approximately 1/15 of the main field area.

(2) Emergence

① To promote elongation of the coleoptile, maintain dark conditions until sprouts grow to about 5~10 mm.

② The optimum temperature for emergence is 32℃ during the day and 30℃ at night, resulting in emergence within about 2 days.

③ If coleoptiles elongate excessively or are exposed to direct sunlight, albino seedlings without chlorophyll are easily produced.

(3) Greening

① Greening is the process of gradually exposing emerged seedlings to light so that chlorophyll is formed and the seedlings are acclimatized.

② A light intensity of 20~30 klux for 1~2 days is suitable for greening.

③ At light intensity above 40 klux with temperatures below 10℃, albino seedlings are easily produced due to stress.

④ Optimum conditions are 25℃ (day) and 20℃ (night), allowing the first and second leaves to develop green coloration.

(4) Hardening

① Hardening is carried out under open-field conditions. For the first 8 days, maintain temperatures of 25℃ (day) and 15℃ (night).

② During the later period (10~25 days), adjust to about 20℃ (day) and 15℃ (night).

③ High humidity, acidic soils, and large temperature fluctuations easily cause damping-off and patchy seedling disorder.

(5) Setting

① Setting refers to placing nursery boxes into the seedbed in the paddy field.

② During setting, keep at least a 10cm distance from vinyl tunnels, and ensure that the holes at the bottom of the boxes are in close contact with the bed soil so that capillary water movement is facilitated.

③ Irrigation should be applied only to the furrows, maintaining the water table 2~3cm below the bottom of the nursery boxes.

④ Complete drainage should be carried out 5~7 days before transplanting.

(6) Fertilization during Nursing

① As basal fertilizer, apply 1~2g of urea per box.

② Depending on seedling growth, at the 3-leaf stage (5~7 days before transplanting), apply 1~2g of urea per box in 100-fold dilution as topdressing.

5) Main Field Preparation

Main field preparation involves plowing, supplying irrigation water, weed removal, leveling, application of compost and fertilizers, and puddling the soil to facilitate transplanting.

(1) Tillage

Tillage is the process of plowing the soil to enhance rice growth while simultaneously suppressing weeds.

Table 6-2. Comparison of characteristics between spring plowing and autumn plowing

Spring plowing	Autumn plowing
① Paddy fields that have undergone autumn plowing must be plowed again in spring. ② In sandy soils, double-cropping fields, and poorly drained paddies, it is preferable to carry out only spring plowing. ③ Spring plowing is commonly practiced in sandy fields, ordinary paddy fields, and poorly drained lowland paddies.	① In dry fields with high organic matter content, autumn plowing is recommended. ② Autumn plowing improves the physico-chemical properties of the soil (dry-soil effect) and also helps control overwintering pests. ③ Autumn plowing is carried out in immature soils, saline-affected fields, and fields where rice straw is applied (even in sandy soils, autumn plowing is practiced if fresh straw is incorporated).

(2) Plowing Depth

① In general, heavy clay loam or clay soils should be plowed deeply, while sandy soils and poorly drained paddy fields should be plowed shallowly.

② The recommended plowing depth is about 18cm, and deeper plowing requires the application of larger amounts of organic matter and fertilizers in order to increase yield.

③ Deep plowing may slightly reduce the initial growth of rice, but since nutrients are supplied until the late growth stage, the number of effective tillers increases, heading is slightly delayed, lodging is reduced, and grain filling improves, thereby increasing yield.

(3) Soil Preparation

① To prevent leakage of irrigation water, the levees should be well maintained, and the field surface should be leveled before applying water and preparing the soil.

② Soil preparation breaks up clods, softens the soil, and helps fertilizers mix uniformly.

③ Making muddy water blocks light penetration, thereby killing weeds or suppressing their growth.

④ In well-drained paddy fields, fine puddling blocks soil pores and prevents leakage.

⑤ In poorly drained paddy fields, coarse puddling facilitates drainage.

⑥ Using a laser land leveler ensures a uniformly flat field surface, which reduces missing hills and patchy seedlings after transplanting, lowers herbicide injury, enhances herbicide efficacy, and allows for more precise crop management.

⑦ A level soil surface promotes good seedling establishment after transplanting.

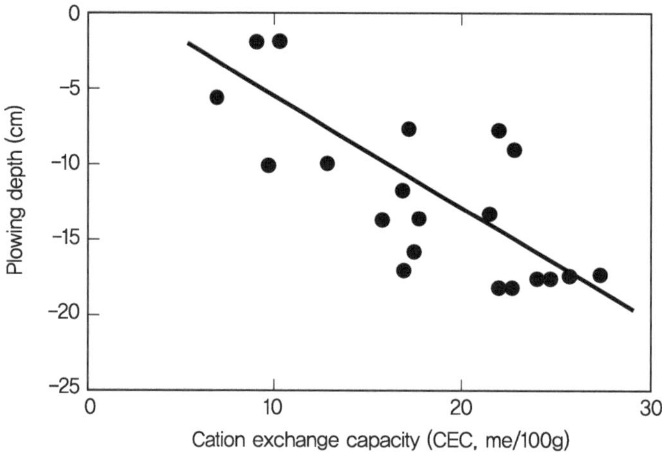

Figure 6-5. Relationship between plowing depth and cation exchange capacity (CEC) (Aomine, 1955).

6) Transplanting

(1) Rice Transplanter

① Transplanting refers to moving seedlings grown in nursery boxes or seedbeds into the main field.

② When transplanting is done too early, low temperatures inhibit growth, extend the vegetative growth period in the main field, increase fertilizer and water consumption, and promote weed infestation.

③ Excessively early transplanting results in overcrowding, many non-productive tillers, poor ventilation, increased incidence of pests and diseases, and greater risk of lodging. In particular, extreme early transplanting leads to grain filling during the hot season, thereby reducing quality and increasing the occurrence of chalky kernels.

④ If transplanting is delayed too much, vegetative growth is insufficient, yields decline, and the occurrence of white-core and white-belly grains increases, thereby lowering rice quality.

(2) Optimum Transplanting Time

① The optimum transplanting time is when heading can occur within the safe heading period, i.e., when panicle emergence and grain filling can be completed before autumn temperatures fall.

② Considering the minimum temperature required for root establishment, the optimum transplanting time is when the soil temperature is at least 11℃ for young seedlings by machine transplanting, 13℃ for medium seedlings by machine transplanting, and 15℃ for mature seedlings by hand transplanting.

③ For the central region of Korea, using medium seedlings of medium–late maturing varieties as a standard, the optimum transplanting time is around May 20.

④ Transplanting with young seedlings delays heading by about 3~5 days compared to medium seedlings, so the transplanting should be done correspondingly earlier.

(3) Safe Heading Limit

① This refers to the heading period in which the average temperature during the 40 days after heading can be maintained at 20~22℃ or higher.

② The optimum transplanting time is determined backward from this safe heading period, adjusted by region and topography.

③ Recently, in addition to yield, transplanting time is also considered to improve grain quality.

(4) Transplanting Methods

A: Planting density and spacing

① For hand transplanting, the standard planting density is three to four seedlings per hill at 30 × 15cm spacing, equivalent to 72 hills per 3.3m².

② For machine transplanting in single-cropping plain areas, 80 hills per 3.3m² are planted, with three seedlings per hill.

③ In double-cropping areas, 85 hills per 3.3m² (four seedlings per hill) are planted, and in mountainous highland or late-planting areas, 120 hills per 3.3m² (six seedlings per hill) are planted.

Table 6-3. Comparison between machine transplanting and hand transplanting

No.	Item	Machine transplanting	Hand transplanting
①	1,000-Grain weight	Smaller	Larger
②	Cultivation management	Somewhat more difficult	Somewhat easier
③	Heading time	Later, this was due to earlier transplanting, and it is safer in single-cropping southern regions	Earlier
④	Labor-saving effect	Large, more than 75% compared with hand transplanting	Small, more labor needed for seedling raising and transplanting
⑤	Land-use efficiency	Lower	Higher
⑥	Minimum establishment temperature	Lower, though excessively early transplanting may cause cold injury	Higher
⑦	Missing hills	More frequent	Fewer, can be adjusted by selecting healthy seedlings
⑧	Number of effective tillers	Greater, because shallow planting places tillering nodes closer to the soil surface	Fewer
⑨	Number of spikelets per panicle	Greater	Fewer
⑩	Number of tillers	More, since tillering starts about one week later and the vegetative period is longer	Fewer
⑪	Optimum transplanting time	Earlier, about one week earlier	Later
⑫	Ripening ratio	Lower	Higher
⑬	Seedling age	About 30 days	About 40 days

B: Planting depth

① The appropriate transplanting depth is 2~3cm, and it is preferable to plant shallowly as long as the seedlings do not fall over.

② Deep planting delays establishment and reduces tillering.

③ Very shallow planting causes seedlings to float or fall, leading to missing hills.

C: Management immediately after transplanting

① Transplanting injury refers to the temporary wilting of seedlings after transplanting because existing roots and root hairs are damaged, losing absorption capacity, until new roots grow and restore water and nutrient uptake.

② To reduce transplanting injury, irrigation should be applied deeply. However, since water applied immediately causes seedlings to float, it is preferable to irrigate about 24 hours after transplanting.

2. Nutrients and Fertilization

1) Nutrients and Fertilization

(1) Inorganic Nutrient Uptake

A: Essential elements in rice

① The macronutrients consist of nine elements, namely C, H, O, N, P, K, Ca, Mg and S, and these are contained in large amounts in the dry matter.

② The micronutrients consist of seven elements, which are Fe, Cu, Zn, Mn, Mo, B and Cl.

③ Except for carbon dioxide (CO_2) supplied from the atmosphere, the other 15 essential elements are absorbed from the soil solution.

④ Although not classified as essential, elements such as silicon (Si), iodine (I), and cobalt (Co) are involved in rice growth. In particular, silicon plays a crucial role in protecting against blast disease, maintaining erect leaf posture, and sustaining the ideal plant type. Therefore, among fertilizers for rice, silicon should be supplied in the largest quantity.

B: Sites and timing of inorganic nutrient uptake

① Nutrient absorption mainly occurs in the 2~3cm region near the root tips. Uptake capacity is stronger in newly formed and healthy roots. For evaluating water and nutrient uptake ability, the number or weight of root tips is more meaningful than total root mass.

② Substances such as oxygen (O_2), carbon dioxide (CO_2), and water (H_2O) are absorbed into the cell by diffusion through the phospholipid layer of the cell membrane.

③ Inorganic nutrients cannot pass through the phospholipid bilayer, and instead they are selectively absorbed with the aid of transport proteins (carrier proteins) located in the cell membrane.

④ The uptake of inorganic nutrients by rice roots varies with the growth stage: nitrogen and potassium are generally absorbed rapidly, whereas phosphorus is absorbed later.

C: Conditions for inorganic nutrient uptake

① Adequate amounts of available nutrients must be present in the soil solution.

② Root tissues must contain sufficient respiratory substrates to provide energy for absorption.

③ Oxygen must be supplied adequately to generate the energy needed for nutrient uptake.

④ Temperature conditions must be suitable for the physiological reactions within the cell.

D: Inhibition of nutrient uptake

① When soils become reduced, gases such as hydrogen sulfide (H_2S) or organic acids such as acetic acid are produced, which interfere with respiratory enzymes, thereby inhibiting the absorption of water and nutrients.

② The optimum water temperature for nutrient absorption by rice roots is 30~32℃. When the water temperature is higher (because the consumption of organic matter and other compounds increases under high temperature) or lower (under low-temperature conditions), the absorption of nutrients such as silicon, phosphorus, potassium, and nitrogen decreases markedly. However, calcium (Ca) and magnesium (Mg) are hardly affected.

③ In summer, when temperature rises, oxygen in paddy soil becomes insufficient, and sulfate (SO_4) is reduced to hydrogen sulfide (H_2S), which inhibits the uptake of inorganic nutrients, with the order of suppression being $P > K > Si > NH_4 > Mn > H_2O > Mg > Ca$.

④ Hydrogen sulfide (H_2S) gas inhibits nutrient absorption by rotting roots, even at very low concentrations of 0.1 ppm.

⑤ When oxygen is deficient, the absorption of potassium (K), manganese (Mn), and silicon (Si) increases because insoluble oxidized forms of these elements become soluble due to soil reduction.

E: Amount and timing of inorganic nutrient uptake

① The uptake of nitrogen, phosphorus, and potassium reaches its maximum about 20~30 days before heading, when the number of new roots is also greatest.

② Nitrogen, phosphorus, and potassium are mainly absorbed by new roots during the vegetative growth stage, whereas phosphorus and potassium are also absorbed by older roots during the reproductive stage.

③ The uptake of iron and magnesium reaches its maximum about 10~20 days before heading.

④ The uptake of silicon and manganese reaches its maximum just before heading.

⑤ The maximum uptake of iron and magnesium, as well as silicon and manganese, occurs 10~20 days later than the period of abundant new roots, but coincides with the time when total root mass is at its maximum.

(2) Inorganic Nutrients and Rice Growth

A: Uptake and translocation of inorganic nutrients

① In rice, the uptake of inorganic nutrients is greatest for silicon (Si), followed by nitrogen (N), potassium (K), phosphorus (P), calcium (Ca), and magnesium (Mg) [Si > N > K > P > Ca > Mg], while their translocation within the plant is highest for phosphorus (P), followed by nitrogen (N), sulfur (S), magnesium (Mg), potassium (K), and calcium (Ca) [P > N > S > Mg > K > Ca].

② Calcium and silicon, which are less mobile, must be supplied in sufficient amounts at the specific growth stages when they are needed.

③ When water temperature is low, nutrient uptake inhibition occurs in the order of Si, P, K, and N.

④ Nutrient uptake inhibition due to root damage caused by hydrogen sulfide (H_2S) decreases in the order of P, K, Si, NH_4^+, Mn, H_2O, Mg, and Ca.

B: Inorganic nutrient uptake according to growth stage of rice

① In inorganic nutrient uptake, nitrogen and phosphorus are absorbed in large amounts during the early growth stage, when protein synthesis is active. Potassium (K), which is involved in photosynthesis and the translocation of photosynthetic products, is absorbed until maturity, while calcium (Ca) is absorbed in large amounts when carbohydrates accumulate and continues to be absorbed until the ripening stage.

② During the reproductive growth stage, the uptake of nitrogen and phosphorus decreases, while the concentrations of calcium, magnesium, and silicon increase.

③ During the vegetative growth stage, rice absorbs large amounts of nitrogen, phosphorus, and potassium, which are used to synthesize proteins and produce stems and leaves.

④ Nitrogen, phosphorus, and sulfur are absorbed in large amounts from early growth to heading, after which nutrients stored in leaves and stems are translocated to panicles, and most assimilates are deposited in the grains.

⑤ It is advantageous to ensure sufficient absorption of nitrogen and phosphorus in the early growth stage so that they can be stored within the plant, while potassium and calcium are continuously absorbed from the early to the middle growth stages.

⑥ Magnesium (Mg) is required in large amounts during panicle initiation, and uptake is highest at this stage.

⑦ In the early growth stage, the concentrations of nitrogen and potassium should be high, whereas in the later growth stage, the concentration of silicon (Si) should be high.

⑧ Nutrient uptake increases until panicle initiation, decreases between panicle initiation and heading, and then declines rapidly after heading.

(3) Physiological Functions and Problems of Excess or Deficiency of Inorganic Nutrients

A: Nitrogen (N)

① Nitrogen is the most abundant element in the atmosphere, accounting for about 79%, and it constitutes the protoplasm components such as DNA, RNA, chlorophyll, and amino acids (proteins). Since it is involved in leaf area expansion and tiller formation, nitrogen deficiency greatly reduces growth and yield.

② Nitrogen content is about 16% of the protein weight. Therefore, the crude protein content can be calculated by analyzing the nitrogen content of the plant and multiplying it by 6.25 (100 ÷ 16).

③ Nitrogen fertilizers dissolve in water and become ammonium (NH_4^+) or nitrate (NO_3^-) ions. In paddy fields, rice mainly absorbs ammonium (NH_4^+) ions and utilizes them for amino acid synthesis. The first amino acid synthesized is glutamine, which is formed by the combination of ammonium (NH_4^+) and glutamic acid. From this glutamine, various other amino acids are synthesized and used in protein synthesis, and glutamine serves as the precursor for nucleic acids and other organic nitrogen compounds.

④ Excess nitrogen results in excessive vegetative growth, reduced light-use efficiency, lower dry matter production efficiency, and weaker plant growth.

⑤ When rainfall is frequent or light intensity is low, nitrogen absorption promotes protein synthesis and vigorous growth of leaves and stems. However, photosynthesis is reduced, resulting in decreased synthesis of cellulose and lignin, the main constituents of cell walls, thereby weakening the plant body.

⑥ When excessive nitrogen fertilizer is applied, or even when the nitrogen application rate is appropriate but photosynthesis is reduced due to insufficient sunlight, relatively high nitrogen remains in the form of ammonium or amino groups, making rice more susceptible to blast disease and other stresses.

⑦ If nitrogen fertilizer is applied in greater amounts than appropriate, rice leaves elongate, leaf area becomes excessively large, and light interception deteriorates, leading to reduced photosynthesis and increased respiration.

⑧ Application of nitrogen fertilizer delays maturation, reduces the harvest index (the ratio of grain to straw), and increases the risk of lodging.

⑨ Short-statured and erect cultivars such as Tongil-type rice are more tolerant to nitrogen fertilizer and generally show stronger resistance to blast disease.

⑩ Application of nitrogen fertilizer increases the protein content of the grain, which reduces eating quality (palatability).

B: Phosphorus (P)

① Phosphorus exists only in small amounts in soil and its concentration is low, yet it is an essential component of genetic and energy-related substances such as DNA, RNA, phospholipids, ATP, NADP, and phytic acid.

② The phosphorus content in crop plants is about 20% of that of nitrogen or potassium, but its importance is very high.

③ Phosphorus is absorbed mainly in the form of $H_2PO_4^-$, rapidly translocated to vigorously growing tissues such as apical meristems, nodes, and panicles, and stored in vacuoles. It also binds with organic matter for metabolic use. In rice seeds, it is stored as phytin (phytic acid salt) and utilized during germination.

④ In flooded paddy soils, the availability of phosphorus increases, so deficiency is less common, but phosphorus application promotes nitrogen uptake and is essential for physiological metabolism.

⑤ In cold regions, phosphorus uptake is further restricted due to low temperatures, so adequate application of phosphate fertilizers is necessary.

⑥ When phosphorus is deficient, leaves become narrow and dark green, heading and grain filling are delayed, plant height is reduced, tillering is poor, and it becomes difficult to secure sufficient effective panicles. Reduced photosynthesis and respiration also lower protein synthesis.

⑦ When excessive phosphorus accumulates in paddy soils, rice plants absorb too much nitrogen while silicon (Si) uptake is inhibited, making them more susceptible to rice blast disease.

⑧ Phosphorus promotes the growth of algae, leading to moss growth in paddy fields and contributing to green or red tides, thereby causing environmental problems.

C: Potassium (K)

① Potassium is one of the nutrients most abundantly absorbed by plants, and it exists as 2.3% of the Earth's crust in primary minerals. Unlike other essential elements, it is not a structural component of cells, but plays a crucial role in regulating physiological metabolism.

② In crop plants, potassium mainly exists as ionizable organic salts, and is involved in enzyme activation, protein synthesis, photosynthesis, translocation of photosynthates, and osmotic regulation.

③ In paddy soils, potassium is sufficiently supplied through irrigation water, so even a basal application alone is usually enough to prevent deficiency symptoms.

④ Potassium has a tendency for excessive uptake, and when applied in large amounts, absorption increases and antagonistic effects may occur, inhibiting the uptake of other ions such as calcium and magnesium.

⑤ When potassium is deficient, protein synthesis is inhibited, carbon assimilation decreases, respiration increases, and as a result, net dry matter production declines.

⑥ Potassium deficiency causes accumulation of ammonium and nitrate nitrogen in plant tissues, producing a physiological state similar to nitrogen over-absorption. This makes plants more susceptible to pests and diseases, reduces cellulose and lignin synthesis, weakens stems, and increases lodging susceptibility.

⑦ Under potassium deficiency, leaf tips and margins become chlorotic, lower leaves wither, and grain setting is poor.

⑧ Potassium is more abundant in the cytoplasm than in vacuoles, and it is also involved in stomatal opening/closing and respiratory activation.

D: Calcium (Ca)

① Calcium constitutes about 3.6% of the Earth's crust. It is a component of the cell membrane, strongly influencing cell division and growth, and together with pectin, it forms the cell wall.

② To increase rice yield, periodic application of lime is required.

③ As a cation, calcium exhibits antagonistic interactions with potassium and magnesium.

④ Since calcium is poorly mobile within the plant body, its deficiency causes the roots, new shoots, and apical meristems of buds to turn reddish and die.

⑤ During panicle initiation and grain filling stages in rice, calcium facilitates the translocation of photosynthates within the plant and also corrects soil acidity.

E: Magnesium (Mg)

① Magnesium is a structural component of chlorophyll, accounting for about 15–20%, directly involved in photosynthesis. It also serves as an activator of several enzymes, including those involved in ATP, CO_2, and glutamine synthesis.

② Magnesium deficiency hampers the development of growing points in stems and roots. Because magnesium from older leaves is translocated to apical meristems or younger leaves, the older leaves undergo chlorophyll degradation, resulting in interveinal chlorosis.

③ As a cation, magnesium shows antagonism with potassium. Rice grains with relatively higher magnesium content (higher Mg/K ratio) are associated with better eating quality (palatability).

④ Along with iron (Fe), magnesium promotes the activity of various photosynthetic enzymes, participates in vitamin synthesis, and plays a role in nitrogen metabolism.

⑤ If magnesium deficiency occurs from the panicle initiation stage to the heading stage, the proportion of sterile spikelets increases, thereby reducing grain yield.

F: Iron (Fe)

① In paddy soils, the solubility of iron increases under flooded and reduced conditions, so deficiency symptoms rarely occur. However, in old paddy soils, iron deficiency may appear due to leaching losses. Deficiency can also occur in poorly drained calcareous areas, because in alkaline soils with high calcium (Ca) content, the solubility of iron and manganese decreases.

② When iron (Fe) is deficient, toxic gases such as hydrogen sulfide (H_2S) are produced, often causing root rot.

③ Iron also shows antagonistic interactions with other cations such as potassium and manganese (Mn).

④ Iron is a component of redox enzymes, including hemoproteins, cytochromes, and ferredoxin, and about 80% of it is present in the chloroplasts.

⑤ Iron is absorbed by the roots and transported to the leaves through the xylem, but its mobility within the plant is poor. Thus, when iron is deficient, young leaves first exhibit reduced chlorophyll content, interveinal chlorosis, and striping symptoms.

G: Silicon (Si)

① Silicon is not considered an essential element for rice, yet rice absorbs and utilizes an amount of silicon about ten times greater than nitrogen. Silicon accounts for 10–15% of rice dry matter, whereas legumes such as soybean accumulate less than 0.5%. Therefore, rice is often referred to as a “silicate plant,” with silicon abundantly deposited in the epidermal tissues of leaves, culms, and husks.

② In soil solution, silicon exists in the form of $Si(OH)_4$, which readily passes through the rice root cell membrane and is transported with water throughout the plant. Deposited silicon in leaves strengthens leaf tissues.

③ In leaves, silicon accumulates beneath the cuticular layer of the epidermis, forming a rigid silica layer, and around the cell membrane, it forms a silica–cellulose composite layer.

④ Silicon accumulation in epidermal cells helps leaves stand erect (erect plant type), improves light interception, reduces cuticular transpiration, prevents water stress, and thereby enhances photosynthesis.

⑤ Silicon strengthens resistance to pests and diseases by making it more difficult for pathogenic fungi such as the rice blast fungus to penetrate.

⑥ Silicon deposition in culms enhances stem strength, allowing them to withstand tension caused by transpiration-induced xylem pressure, promotes aerenchyma development, and improves lodging resistance.

⑦ Silicon complements nitrogen, and when the silicon-to-nitrogen ratio is high, rice plants grow more vigorously. Even under excessive nitrogen fertilization, rice with high silicon uptake grows sturdily, whereas under high-nitrogen cultivation, silicon deficiency is more likely to occur.

⑧ Although soils generally contain abundant silicon oxides, the amount of plant-available soluble silicic acid is often insufficient, making separate application of silicate fertilizers necessary.

⑨ Application of silicate fertilizers enhances the availability of phosphate, whereas deficiency results in drooping and wilting leaves, necrotic spots, and susceptibility to iron and manganese toxicities.

2) Nutritional Disorders

(1) General symptoms of nutritional disorders in rice

A: Deficiency or excess of mineral nutrients

① Abnormal plant height (excessive elongation or stunting) and a reduction in tiller number occur when mineral nutrients are deficient or excessive.

② Leaf color changes to chlorotic, brownish, or orange. Interveinal chlorosis appears when potassium or magnesium is deficient, while uniform chlorosis of the entire leaf occurs when nitrogen or sulfur is deficient.

B: Mobility of mineral nutrients

① Deficiency symptoms of mobile nutrients such as nitrogen (N), phosphorus (P), potassium (K), and sulfur (S) first appear on the lower leaves, since nutrients in older leaves readily translocate to younger leaves.

② Deficiency symptoms of immobile nutrients such as iron (Fe), boron (B), and calcium (Ca) appear on the upper leaves, since nutrients in lower leaves cannot easily move upward.

③ When mineral nutrients are excessively absorbed, they tend to accumulate in the lower leaves, so toxicity symptoms usually appear first in the lower leaves.

④ Under strongly reduced soil conditions in paddy fields, rice roots turn black with a foul odor, and hydrogen sulfide (H_2S) toxicity occurs. Normally, newly grown healthy roots are whitish, whereas older roots appear brown.

(2) Deficiency Symptoms by Element

① Nitrogen deficiency suppresses tillering and growth, producing narrow, short, erect leaves that turn pale green, while the older leaves wither and die.

② Combined deficiency of phosphorus and potassium results in stunted growth, leaves turning dark green, drying of older leaves, and reduced tillering.

③ Magnesium deficiency causes leaves to droop, beginning with interveinal chlorosis that spreads until the entire leaf turns yellow.

④ Zinc (Zn) deficiency produces whitening of the midrib in young leaves, while older leaves develop brown spots and striping. Tillering continues, but plant height is reduced, growth becomes uneven, and maturity is delayed.

⑤ Iron (Fe) deficiency often causes root rot and whitening of leaves, but these symptoms disappear once iron is supplied.

(3) Diagnosis of Rice Nutritional Status: Diagnosis Based on Leaf Color

① Nutritional diagnosis by the iodine-staining reaction of the leaf sheath: the greater the iodine reaction value, the higher the starch content in the plant.

② Lower iodine reaction values indicate higher internal nitrogen content, and plants become more susceptible to rice blast and bacterial leaf blight.

③ Long but narrow leaf blades indicate silicon deficiency caused by excess nitrogen.

④ Short and narrow, firm, pale-green leaves indicate nitrogen deficiency, whereas short but wide, dark-green leaves indicate potassium deficiency.

Table 6-4. Nutrient deficiency symptoms in rice

Fertilizer classification	Nutrient	Affected plant part	Plant height	Leaf color	Leaf shape	Root	Tillering	Maturity	Disease occurrence	Others
Macro nutrients	N	Old leaves	Growth inhibited	Pale green	Narrow, short, erect	–	Suppressed	Accelerated	Increased	Whole field shows chlorosis
	P			Dark green	Narrow, erect	–		Delayed	–	Older leaves turn dark brown
	K		Dwarf	Dark green, Yellow stripes	Rusty-brown spots, leaf rolling	Weakened	–	Early maturity	Increased	Early wilting, Late-stage deficiency symptoms
	Mg		–	Intermittent green, Interveinal chlorosis	entire leaf bleaching		–	–	–	Easily translocated through phloem
	S		Growth reduced	Chlorosis starting from upper leaves, Pale green, Bleaching	–	–	Tillering reduced	Delayed	–	–
	Ca	Young leaves	–	Leaf tip chlorosis	necrosis, tearing	Weakened	–	–	–	Rare in lowland rice
Micro nutrients	Fe		–	Chlorosis of leaf veins in young leaves	reduced chlorophyll	–	–	–	–	Occurs in upland rice, rare in paddy rice
	Mn		Dwarf	Young leaf tips turn grayish-green, Chlorotic, bleached	Necrotic spots	–	–	–	–	Occurs in upland rice, rare in paddy rice
	Cu		–	Yellow-striped leaves with bluish-green coloration	Wilting of young leaves	–	Tillering reduced	–	–	Sterility increases
	B		Dwarf	White tips of young leaves	Leaf rolling	–	–	–	–	Poor panicle development, rare in paddy rice
	Zn		Growth suppressed		Weak, drooping leaves	–	Tillering inhibited	–	–	Sparse and uneven growth
Special element	Si	Whole leaf	–	–	Weak, drooping	–	–	–	Increased	Increased lodging

Table 6-5. Critical concentrations of mineral nutrients (Deficiency/Excess) in the upper leaves of rice plants at the panicle differentiation stage

Fertilizer classification	Nutrient	Optimal concentration	Deficiency level	Toxicity level
Macronutrients	Nitrogen (N)	2.9–4.2%	0.25%	0.45%
	Potassium (K)	1.8–2.6%	1.5%	2.0%
	Calcium (Ca)	0.3–0.6%	0.1%	0.5%
	Phosphorus (P)	0.2–0.4%	0.15%	0.7%
	Magnesium (Mg)	0.17–0.3%	0.1%	0.5%
	Sulfur (S)	0.15–0.25%	0.1%	0.5%
Micronutrients	Iron (Fe)	75–150 ppm	60 ppm	300 ppm
	Manganese (Mn)	50–500 ppm	30 ppm	800 ppm
	Zinc (Zn)	25–50 ppm	20 ppm	500 ppm
	Copper (Cu)	7–15 ppm	5 ppm	25 ppm
	Boron (B)	6–15 ppm	5 ppm	100 ppm
Special element	Silicon (Si)	8–10%	5%	–

* 1% = 10,000 ppm

3) Fertilization

(1) Determination of Fertilizer Application Rate

The amount of fertilizer to be applied is calculated by subtracting the naturally supplied nutrients from the nutrient requirement for the target yield, and then dividing by the absorption efficiency of the applied fertilizer.

In formula form:

$$\text{Fertilizer absorption rate} = \frac{\text{Nutrient requirement} - \text{Natural supply}}{\text{Absorption efficiency of fertilizer nutrien}}$$

For instance, if the nutrient requirement of rice is 14 kg, the natural nutrient supply is 4 kg, and the fertilizer absorption efficiency is 50%, the required fertilizer nutrient amount is calculated as:

Required nutrient amount = (14–4) / 0.5 = 20kg

Since urea contains 46% nitrogen, the actual amount of urea required is calculated by dividing by its effective nitrogen content:

Required urea application = 20 / 0.46 ≈ 43kg

Thus, approximately 43kg of urea is required.

For example, if the nutrient requirement of rice is 14kg, the natural nutrient supply is 4kg, and the absorption efficiency of fertilizer is 50%, the required amount of available nitrogen becomes (14–4)/0.5=20kg. Since urea contains 46% nitrogen, the actual application rate of urea is calculated by dividing 20 by 0.46, which results in approximately 43kg. Therefore, about 43kg of urea is required.

(2) Timing of Fertilizer Application

① Fertilizer should be applied in split doses according to the growth period and developmental stage of rice to enhance fertilizer use efficiency and promote proper plant growth.

② Basal fertilizer is applied before transplanting, and once seedlings are established, the nutrients are absorbed immediately and promote tillering, thereby ensuring a sufficient stem number.

③ Tiller fertilizer is applied 12~14 days after transplanting. If transplanting is delayed, the amount of tiller fertilizer is reduced.

④ Panicle fertilizer (also called heading fertilizer) is applied about 24~25 days before heading, when the panicle primordium reaches 1~1.5mm in length, to increase the number of spikelets per panicle.

⑤ Grain fertilizer (also called ripening fertilizer) is applied at heading to increase grain weight, usually at about 10% of the total nitrogen application.

⑥ Grain fertilizer is most effective when the nitrogen content of the functional leaves is below 2%, but it should not be applied if nitrogen levels are already high or under poor light conditions or low temperatures.

⑦ When the objective is to produce high-quality rice, it is recommended not to apply nitrogen as grain fertilizer.

(3) Recommended Fertilizer Standards

① In typical paddy fields with machine transplanting, the recommended application rate is N–P–K at 9–4.5–5.7 kg per 10a.

② Fertilizer recommendations are closely related to rice yield potential, and until the 1980s, application rates were increased to maximize yield.

③ More recently, with rice self-sufficiency and greater emphasis on eco-friendly farming and grain quality, nitrogen application rates have been reduced. To produce rice with good eating quality, nitrogen application should be reduced below 9 kg/10a, which lowers grain protein content (below 6%) and improves palatability.

(4) Split Application and Fertilizer Ratios

① Nitrogen fertilizer is generally split-applied, phosphorus fertilizer is applied entirely as basal fertilizer, and potassium fertilizer is divided between basal and panicle fertilizer at a 7:3 ratio.

② Under timely transplanting in plain areas, nitrogen fertilizer is split in the ratio of 50:20:30 for basal, tiller, and panicle fertilizer, respectively.

③ Grain fertilizer with nitrogen is favorable for yield increase, but since it also raises grain protein content and reduces eating quality, the amount should be minimized.

④ In deepwater-type varieties, basal fertilizer is increased.

⑤ In early-season cultivation, where the growth duration is extended, the split application amount should be increased accordingly.

⑥ In sandy soils or leaky paddy fields, basal fertilizer should be reduced and split application should be increased.

⑦ Under favorable climatic conditions with strong photosynthetic activity, topdressing should be increased to improve yield.

(5) Rational Fertilization

① Rational fertilization means applying appropriate fertilizer amounts based on soil testing results and official fertilization recommendations.

② To increase yield, it is important to secure sufficient leaf area and panicle number, while avoiding lodging.

③ Grain protein content should be reduced to improve eating quality. To achieve this, the timing and amount of nitrogen application should be managed so that excess nitrogen is not translocated into grains after heading.

④ Nitrogen availability should be maintained until the effective tillering stage, and fertilizer application should be completed before the panicle differentiation stage, about 30 days prior to heading.

⑤ Panicle fertilizer should be managed carefully during elongation of the upper 4~5 internodes to avoid excessive nitrogen, which can increase lodging risk.

⑥ In rice cultivation, grain fertilizer can help secure high yields if panicle fertilizer was insufficient, but nitrogen grain fertilizer tends to increase grain protein content and lower eating quality, so it should not be applied when the goal is to produce high-quality rice.

⑦ In paddies irrigated with cold water, additional phosphorus and potassium fertilizer should be applied.

⑧ In fields improved through soil dressing, land consolidation, or deep plowing, nitrogen, phosphorus, and potassium should be increased by 20~30% to secure yield.

⑨ In fields with short sunlight duration, or prone to cold damage, flooding, and lodging, nitrogen should be reduced while phosphorus and potassium should be increased by 20~30%.

(6) Fertilizer Types and Application Methods

① Single-nutrient fertilizers contain only one nutrient element. Urea is used for nitrogen, fused superphosphate for phosphorus, and potassium chloride for potassium.

② Since applying the three nutrients separately is cumbersome, compound fertilizers containing all three macronutrients are used to reduce the number of applications.

③ Most single-nutrient fertilizers are fast-acting and release nutrients immediately after application, which may not coincide with rice nutrient demand. Therefore, slow-release fertilizers are formulated to match nutrient uptake during the growth period.

④ Slow-release compound fertilizers reduce both the frequency and amount of application, are convenient to use, and maintain effectiveness for up to three months, making additional topdressing unnecessary.

⑤ Customized blended fertilizers (BB: bulk blending fertilizers) are produced based on soil test results to provide the precise nutrient composition required for rice growth.

Table 6-6. Types of fertilizers for paddy rice, Their chemical formulas, and nutrient content

Nutrient	Form	Fertilizer type	Chemical formula	Nutrient content(%)
Nitrogen (N)	Urea form	Urea	$(NH_2)_2CO$	46
	Nitrate form	Calcium nitrate	$Ca(NO_3)_2 \cdot 4H_2O$	10
		Chile saltpeter (Sodium nitrate)	$NaNO_3$	20
	Ammonium form	Ammonium sulfate	$(NH_4)_2SO_4$	21
		Ammonium chloride	NH_4Cl	25
		Ammonium nitrate	NH_4NO_3	33
Phosphorus (P)	Water-soluble	Single superphosphate	$Ca(H_2PO_4)_2 \cdot H_2O$	20
		Triple superphosphate	$Ca(H_2PO_4)_2$	46
	Citrate-soluble	Fused magnesium phosphate	$MgO_3 \cdot CaO \cdot P_2O_5 \cdot 2SiO_2$	20
		Mixed phosphate fertilizer	Fused phosphate+ Superphosphate	17
Potassium (K)	Water-soluble	Potassium chloride (Muriate of potash)	KCl	60
		Potassium sulfate	K_2SO_4	50
Calcium (Ca)	–	Quick lime	CaO	80
		Slaked lime	$Ca(OH)_2$	60
		Limestone (Calcium carbonate)	$CaCO_3$	50

* If 7kg of nitrogen is to be applied per 10 ares of paddy field using ammonium sulfate $(NH_4)_2SO_4$: Ammonium sulfate contains 21% N. Required fertilizer amount = 7/0.21 = 33.3 kg.

3. Water Management in Paddy Fields

1) Irrigation and Water Requirement

① Water requirement is defined as the total amount of water absorbed during the growth period divided by the total dry matter, indicating the amount of water needed to produce 1g of dry matter.

② The water requirement of lowland rice (paddy rice) is about 250~300, while that of upland rice and soybean is about 300~450. Although rice requires large amounts of water as an environmental factor, the portion directly used for biomass production is relatively small.

③ Rice is grown in flooded fields not because of a strict physiological water demand, but mainly for agronomic convenience and the beneficial environmental functions of water.

④ Water is essential for rice physiology, functioning in nutrient and water transport, transpiration, photosynthesis, and other metabolic processes.

⑤ In rice cultivation, most water serves as environmental water, favorably influencing growth by regulating nutrient supply and uptake, redox balance of paddy soils, soil and water temperature, removal of harmful substances, suppression of weed emergence, reduction of pest and disease damage, and mitigation of salt accumulation in reclaimed soils.

2) Effects of Irrigation

① Irrigation regulates rice growth and paddy soil conditions, thereby supplying nutrients, promoting growth, and increasing yield.

② It stabilizes water temperature and soil temperature in paddy fields.

③ It promotes soil reduction, preventing excessive decomposition of soil organic matter.

④ It softens the soil, making tillage, puddling, and weeding easier.

⑤ It prevents pests and diseases and suppresses weed emergence.

3) Water Requirement and Irrigation Requirement

(1) Water Requirement (Irrigation Requirement)

① The irrigation water requirement is the total volume of water needed for rice cultivation, i.e., the sum of all water consumed during the growth period.

② It is calculated as:

Irrigation Requirement = Leaf transpiration + Water surface evaporation + Percolation loss + Runoff and seepage

③ Since water is lost by percolation and seepage, the actual irrigation requirement is larger than the basic crop water requirement.

④ he irrigation water requirement is greatest at the panicle initiation stage, followed by the panicle development and transplant establishment stages, then the heading and flowering stage, and is lowest at the ineffective tillering stage.

⑤ Depending on cultivation method, water requirement differs: compared to mechanical transplanting, tilled direct seeding requires about 40% more, and no-tillage direct seeding about 60% more.

Example : If leaf transpiration = 500mm, water surface evaporation = 400mm, and percolation loss = 650mm, then:
Total Water Requirement = 500 + 400 + 650 = 1,550mm
Since runoff is negligible, it can be omitted.

(2) Irrigation Requirement

① The actual irrigation requirement is the deficit obtained by subtracting effective rainfall from the total water requirement.

② In Korea, effective rainfall is assumed to be 70% of average rainfall during the irrigation period.

③ The drainage/percolation rate of Korean paddy soils is about 20~30mm per day.

④ Example: If the total water requirement is 1,500kℓ and the effective rainfall is 450kℓ, then the irrigation requirement is obtained by subtracting effective rainfall from the total water requirement.
Irrigation Requirement = 1,500 − 450 = 1,050kℓ

⑤ Example: If leaf transpiration is 600mm, surface evaporation is 400mm, percolation is 500mm, and effective rainfall is 300mm, then the irrigation requirement is obtained by subtracting effective rainfall from the sum of these values, that is (600+400+500) − 300 = 1,200mm.

4) Water Management in the Main Field

(1) Transplanting Stage

① During transplanting, the water depth should be kept shallow, about 2~3cm, to facilitate the operation.

② If the water depth is too great, seedlings cannot be transplanted properly, and the planted seedlings may float, leading to missing hills.

③ If the water depth is too shallow, transplanting itself becomes easier, but since irrigation cannot be applied immediately after transplanting, seedlings may suffer from severe transplanting shock.

(2) Establishment Stage (Rooting Stage)

① For about 7~10 days after transplanting, irrigation should be maintained at a depth of 6~10cm to prevent transplanting shock.

② Seedlings should be submerged to about one-third to one-half of their height so that transpiration is suppressed and transplanting shock is minimized.

③ When cold water irrigation is unavoidable, the irrigation water should be diverted or dispersed before entering the field.

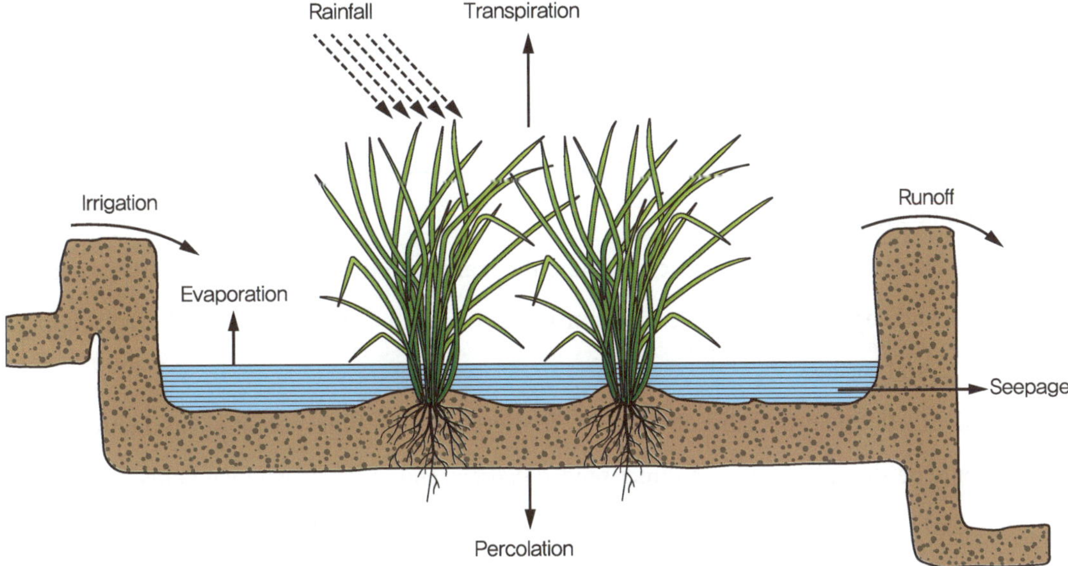

Figure 6-6. Water balance in paddy rice cultivation (Trewartha, 1968)

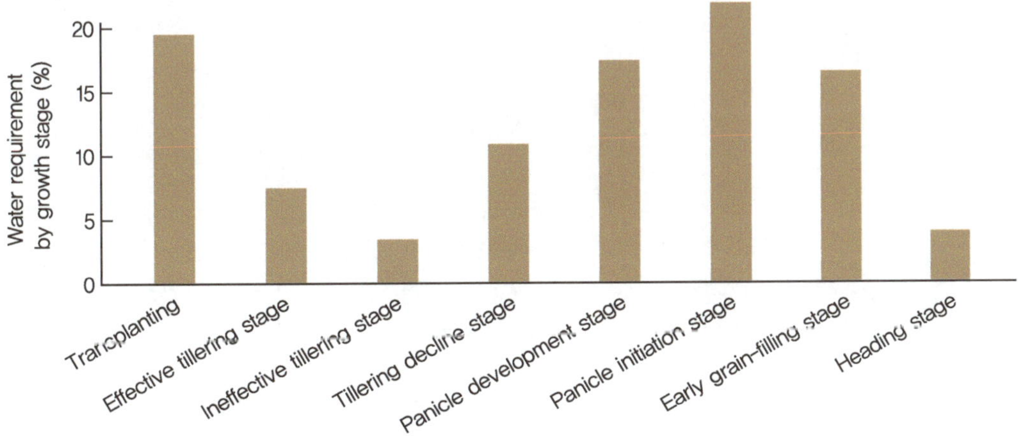

Figure 6-7. Water requirement of rice at different growth stages (Ito, 1962)

(3) Tillering Stage

① During the tillering stage, irrigation should be shallow, about 1~3cm in depth, to promote tiller production.

② A shallow water depth allows the temperature near the growing point to rise during the day under sunlight and to fall at night, and this diurnal temperature fluctuation stimulates tillering.

③ n cold regions or in areas where chilling injury is likely, deep-water irrigation is applied to retain warmth and raise the water temperature.

(4) Ineffective Tillering Stage and Midsummer Drainage

① Around 35 days before heading, during the ineffective tillering stage, irrigation should be withheld (midsummer drainage) to suppress unnecessary tillers, promote aeration, and stimulate growth.

② Midsummer drainage is practiced around the maximum tillering stage to maintain healthy roots and suppress ineffective tillers, usually for about 7 days, until small cracks appear on the soil surface.

③ This practice is especially effective in fertile soils with high potential productivity, and should be strongly applied in direct-seeded paddies and overly vegetative fields. In contrast, in sandy soils or saline soils with poor growth, it should be omitted or applied lightly.

Figure 6-8. Photograph of midsummer drainage

④ Midsummer drainage prevents excessive nitrogen uptake, suppresses ineffective tillering, supplies oxygen to the soil, and removes harmful gases such as hydrogen sulfide, thereby enhancing root activity and preventing root rot. It also promotes deeper rooting, improves nutrient uptake during later growth stages, and enhances grain filling.

⑤ Midsummer drainage hardens the soil surface and strengthens the basal nodes, preventing lodging. When soil becomes oxidized, the potassium-to-nitrogen ratio (K/N) in rice tissues increases, strengthening plant structure and reducing the incidence of diseases such as sheath blight.

⑥ However, midsummer drainage promotes oxidation of ammonium nitrogen in the soil into nitrate, leading to denitrification and reduced fertilizer efficiency.

(5) Panicle Development and Heading Stage

① From the panicle development stage through the panicle initiation stage to the heading stage, rice plants have the largest leaf area, the highest transpiration rate, and the greatest water demand. This is also the most sensitive period to environmental stress.

② Intermittent irrigation is practiced by maintaining a water depth of 2~4cm for a few days, then allowing the field to dry before re-irrigating. This suppresses ineffective tillering and saves water.

③ Continuous flooding is not necessary as long as water is not deficient, and after midsummer drainage, intermittent irrigation supplies oxygen to the root zone and maintains healthy root growth.

④ Since the period of panicle development, flowering, and fertilization is very sensitive to water stress, sufficient irrigation must be ensured.

⑤ A modified form of intermittent irrigation known as alternate wetting and drying (AWD) involves irrigating for 3 days followed by drainage for 2 days. This enhances root activity and removes harmful substances such as methane, hydrogen sulfide, and organic acids in the soil, thereby promoting panicle growth and development.

Table 6-7. Water management by growth stage of rice

No.	Growth stage	Water management method	Water depth (cm)	Effect
①	Transplanting	Shallow irrigation	2~3	Stable transplanting by planting seedlings shallowly
②	Establishment	Deep irrigation	5~7	Reduces transplanting shock, suppresses transpiration, promotes rooting
③	Effective tillering	Shallow irrigation	2~3	Promotes tillering
④	Ineffective tillering	Midsummer drainage (about 1 week)	0	Suppresses ineffective tillers, removes harmful substances, prevents lodging
⑤	Panicle development	Intermittent irrigation (3 days flooded, 2 days drained)	2~4	Enhances root activity, removes harmful substances
⑥	Heading	Normal irrigation	3~4	Promotes pollination and fertilization
⑦	Grain-filling	Shallow intermittent irrigation	3~4	Promotes grain filling, maintains root activity, removes harmful substances
⑧	Maturity/Harvest	Complete drainage (about 1 month after heading)	0	Improves grain quality, facilitates harvesting

(6) Grain-Filling Stage

① Water is still essential during the grain-filling stage, but since leaf transpiration and water-surface evaporation decrease significantly at this time, deep-water irrigation is not necessary. To facilitate nutrient translocation and accumulation, shallow irrigation or alternate wetting and drying is practiced.

② Intermittent irrigation should be applied to ensure adequate oxygen supply to the root zone.

③ Irrigation must be maintained until about 30 days after heading to improve grain quality.

④ If drainage is conducted too early in order to facilitate combine harvesting, grain weight decreases, and the proportion of chalky or immature grains increases, resulting in both lower yield and inferior quality. Early drainage can increase the proportion of immature or chalky grains by 20~30%.

Figure 6-9.
Drainage in a paddy field cultivated with Samkwang just before harvest.

5) Harvest

(1) Optimum Harvesting Time

① The proper harvest stage is when the panicle neck loses its green color and turns yellow, and more than 90% of the grains have matured. At this time, the moisture content of brown rice is about 25%.

② Grain development is completed about 35~40 days after heading, making harvest possible. However, considering yield, milling quality, grain appearance, and starch properties, the optimum harvest time is 40~50 days after heading, when brown rice development is complete.

③ It is recommended to harvest extremely early-maturing cultivars about 35 days after heading, early-maturing cultivars at 40 days, medium-maturing cultivars at 45 days, and medium-late maturing cultivars as well as late-planted rice around 50 days after heading.

④ Grains intended for seed should be harvested about 5 days earlier than those for consumption.

(2) Determination of the Harvesting Time

① The number of days from heading to harvest is primarily determined by accumulated temperature. The effective of accumulated temperature required for grain filling is about 1,000~1,100℃.

② High temperatures during grain filling accelerate the harvest time, while low temperatures delay it.

③ Excessive heat during grain filling increases the proportion of white-back kernels, basal-white kernels, and milky-white kernels. Low-temperature conditions increase immature kernels, chalky kernels, and double-white kernels.

④ Considering that autumn temperatures have recently risen by 2~3 ℃ due to global warming, slightly earlier harvesting is advantageous for producing rice with better eating quality.

(3) Changes in Grain Quality according to Harvesting Time

① Harvesting at the optimum time produces glossy grains with soft stickiness and good aroma. If harvesting is delayed, gloss, aroma, and stickiness decrease, reducing eating quality.

② If harvesting is too early, grain gloss is good, but the incidence of green kernels, partially filled kernels, and broken kernels increases.

③ If harvesting is too late, transparency and gloss decrease, while white-back kernels, damaged kernels, and chalky kernels increase. The bran layer also becomes thicker, shattering losses increase, and damage from birds and rodents becomes more likely.

④ If harvesting is delayed and lodging occurs, yield decreases and harvesting requires much more labor.

(4) Preparation for Optimum Harvesting

① Recognize the importance of harvesting time and secure combine harvesters and operators in advance.

② Freshly harvested paddy rice has a high moisture content of about 25%, which requires a long drying time, and therefore dryers should be prepared.

③ In poorly drained fields or small fragmented paddies where harvesting is easily delayed, farmland consolidation should be undertaken.

④ For large-scale cultivation, it is desirable to stagger transplanting dates or arrange cultivars with different maturation periods to distribute labor demand.

(5) Harvesting Methods Transition

① For seed grains, threshing should be performed slowly at about 300 rpm to avoid damage, which ensures high germination rates and reduces disease incidence.

② Along with combine harvesters, the two-row binder (reaper-binder) was introduced in 1977.

③ Combines, which perform cutting and threshing simultaneously, became widespread in the mid-1980s.

④ Today, the mechanization rate of harvesting is nearly 100%. Compared with manual harvesting using sickles, work efficiency is about 13 times higher with binders and 50 times higher with combines.

⑤ In power-driven threshers, excessively high rotational speed increases efficiency but results in grain damage and broken kernels, so an appropriate speed is around 500 rpm.

⑥ Because high moisture content promotes microbial growth, component migration, and enzymatic activity, harvested rice should be dried immediately to maintain quality.

Rice Crop Science

Chapter 7. Rice Protection

Chapter 7.
Rice Protection

The history of agriculture can be described as a constant struggle against weeds, diseases, and insect pests. Weather-related disasters, on the other hand, were generally expected according to climate conditions, and farmers developed various cultivation techniques to cope with them. Today, diverse and powerful pesticides (crop protection chemicals) have been developed, and agricultural inputs such as fertilizers are abundantly available. However, a high level of understanding of these factors is essential for efficient and integrated pest management. In this chapter, we will study the concepts and practices for controlling and managing these factors.

1. Control of Rice Diseases

1) Classification of Rice Diseases

Table 7-1. Major diseases of rice

<table>
<tr><th>No.</th><th>Disease name</th><th>Pathogen</th><th>Transmission (vector)</th><th>Overwintering form/site</th><th>Characteristics</th></tr>
<tr><td>①</td><td>Brown spot</td><td rowspan="2">Fungus (Ascomycota)</td><td rowspan="3">Wind (seed-borne)</td><td>Infected rice straw / seeds</td><td>Occurs in sandy soils and old paddy fields</td></tr>
<tr><td>②</td><td>Bakanae disease</td><td>Seed surface</td><td>Produces gibberellins</td></tr>
<tr><td>③</td><td>Rice blast</td><td>Fungus (Deuteromycota / Ascomycota anamorph)</td><td>Rice straw / infected seeds</td><td>It favors low temperature and high humidity, and is influenced by silicon fertilization</td></tr>
<tr><td>④</td><td>Sheath blight</td><td>Fungus (Basidiomycota)</td><td>Water</td><td>Soil surface (sclerotia and basidiospores)</td><td>Forms sclerotia and basidiospores</td></tr>
<tr><td>⑤</td><td>Seedling blight</td><td>Fungus</td><td>Water, soil</td><td>Infected tissues / soil</td><td>Frequently occurs in seedling nursery boxes</td></tr>
</table>

⑥	Bacterial seedling rot	Bacterium	Water (seed-borne)	Seeds	Seed-borne transmission
⑦	Bacterial leaf blight		Water	Weeds (e.g., Leersia spp.); rice stubble	Outbreaks after typhoon and flooding
⑧	Rice black-streaked dwarf disease	Virus	Small brown planthopper (Laodelphax striatellus)	Vector insects (weeds, wheat, milk vetch)	Nymphal and transovarial transmission
⑨	Rice dwarf disease		Green rice leafhopper (Nephotettix cincticeps), zigzag leafhopper (Recilia dorsalis)		The pathogen is transmitted during both nymphal and adult stages, and transmission also occurs transovarially
⑩	Rice stripe virus disease		Small brown planthopper (Laodelphax striatellus)		Transmission occurs during the nymphal stage, and it also takes place through transovarial transmission

2) Methods of Rice Disease Control

Table 7-2. Methods of pest control (including insect pests and weeds)

No.	Category	Major control methods
①	Physical control	• Control by mechanical means, or by manipulating temperature, humidity, and light (e.g., insect traps using UV or fluorescent lamps). • Hot-water seed disinfection, destruction of pest habitats by burning collected host materials. • Installation of sticky traps (yellow, blue, red panels) or yellow water pans. • Use of insect-proof nets in greenhouses and nurseries (covering ventilation windows and doors) to block aphids, moths, and other pests. • Bagging of fruits to prevent insect damage. • Soil sterilization by steaming, flooding, or solarization before transplanting. • Cutting and burning of heavily infested plant parts.

②	Chemical control (pesticides)	• Application of pesticides, mainly preventive and at the optimum timing. • Appropriate dilution, dosage, and frequency of spraying to prevent overuse. • Apply sufficient amounts evenly across the field (avoid phytotoxicity and uneven spray marks). • Alternate different pesticide formulations to delay resistance development. • Disadvantages include disruption of ecological balance, resistance induction, and environmental pollution, while advantages are rapid and reliable effects. • Pheromones and microbial pesticides are also included in this category.
③	Cultural control	• Use of cultivation practices such as tillage, adjustment of planting dates, and transplanting of healthy, disease-free seedlings. • Cultivation of resistant varieties. • Crop rotation and intercropping are practiced to eliminate pathogen hosts, and trap crops or antagonistic plants are used to reduce pest populations. • Sanitation of fields to remove inoculum sources: collection and burning of crop residues, elimination of overwintering sites (fallen leaves, pruned branches), and control of alternative hosts (weeds and intermediate hosts). • Disinfection of pruning tools and nursery equipment to prevent pathogen spread.
④	Biological control	• Use of natural enemies or antagonistic organisms (e.g., predators, parasitoids, entomopathogenic microorganisms). • Introduction of antagonistic crops to suppress pests. • Biological control is relatively slow but environmentally safe and sustainable.
⑤	Integrated Pest Management (IPM)	• Concept introduced due to increased environmental burden from heavy use of fertilizers and pesticides. • Optimal and minimal use of pesticides (including biological control agents) in a regionally efficient manner. • Application of the most suitable and efficient methods with minimal environmental impact, rather than indiscriminate combination of all methods.

3) Diseases of Rice

(1) Brown Spot (Cochliobolus Leaf Spot)

① Causal organism is *Cochliobolus miyabeanus*, fungus (Ascomycota).

② The pathogen overwinters as mycelium on rice straw and produces conidia in spring, which are dispersed by wind.

③ Elliptical dark-brown lesions appear on leaves, and the panicle neck, branches, and grains turn dark brown.

④ The disease is prevalent under hot and humid conditions, in acidic sandy soils, and when nutrient deficiency occurs.

⑤ Control measures include cultivation of resistant varieties, seed disinfection, soil amendment, and appropriate fertilizer application.

(2) Bakanae Disease (Foolish Seedling Disease)

① Causal organism is *Gibberella fujikuroi*, fungus (Ascomycota).

Figure 7-1. Lesion symptoms and pathogen of rice bakanae disease (National Crop Pest Management System)

② The pathogen survives as conidia on the seed surface and develops well under high temperature.

③ It occurs severely in acidic soils, is transmitted through seeds, and causes abnormal elongation due to gibberellin activity.

④ Diseased seedlings grow pale green, slender, and abnormally elongated, often more than twice the size of healthy seedlings, and eventually die.

⑤ Recommended measures are to select resistant varieties, use healthy seeds, and apply seed disinfection.

⑥ Diseased seedlings should be removed during the nursery stage or at the early growth stage in the main field.

Figure 7-2. Symptoms of bakanae disease in rice (National Crop Pest Management System)

(3) Rice Blast Disease (Blast)

① Causal organism is *Pyricularia oryzae*, fungus (Deuteromycota).

② This is one of the most destructive diseases of rice in Korea and throughout Southeast Asia. Excessive application of nitrogen fertilizer promotes outbreaks. It can infect all rice organs and consists of diverse pathogen races.

③ The pathogen overwinters as mycelia or conidia on infected straw or seeds. Conidia are easily dispersed by wind and typically have two septa.

④ For infection to occur, airborne conidia require moderately low temperatures of 25~28°C and nearly 100% relative humidity.

⑤ Outbreaks are frequent under conditions of cool, wet weather followed by rapid temperature rise, late transplanting, excessive nitrogen supply, and high humidity. Disease also develops when water is insufficient during the growing season, when cold water is deeply irrigated, when topsoil is shallow, or when fertilizer uptake is sudden and excessive.

⑥ The disease occurs on all parts of the rice plant from seed to panicle, producing spindle-shaped brown lesions. Speck-type lesions form between veins and do not produce conidia.

⑦ White-spot lesions occur on seedlings under dry soil and high-fertility conditions and can produce conidia.

⑧ Acute lesions develop under unfavorable weather or with excessive nitrogen, when host resistance is weakened. These lesions are dark green, elliptical or irregular, and a single lesion can produce 2,000~3,000 conidia.

⑨ Chronic lesions are typical spindle-shaped spots with brown margins and grayish-white centers, where abundant conidia formation and dispersal occur.

⑩ Control methods include cultivation of resistant varieties, avoidance of excessive nitrogen fertilization, warming of cold irrigation water before use, and seed disinfection with fungicides. Additional measures include removal of diseased plants, restriction of nitrogen over-application, and balanced fertilization including silicon (Si).

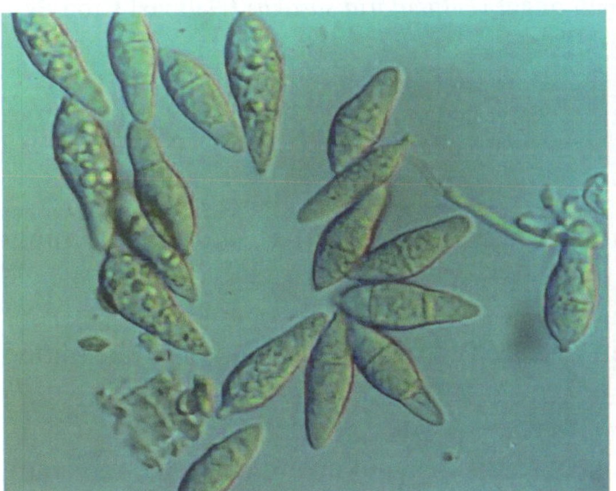

Figure 7-3. Pathogen and lesion symptoms of rice blast (National Crop Pest Management System)

4) Sheath Blight

① Causal organism is *Pellicularia sasakii*, fungus (Basidiomycota).

② The pathogen forms sclerotia near lesions, which are initially white and gradually turn light brown.

③ The disease develops under moderately low temperatures of 23~25°C, nearly 100% relative humidity, excessive nitrogen fertilization, and dense planting conditions.

④ Symptoms appear as irregular dark grayish lesions on the sheath surface.

⑤ Control measures include puddling before transplanting to reduce sclerotia, avoidance of dense planting and excessive nitrogen application, and increased application of potassium fertilizer.

⑥ After the rainy season, when hot summer weather continues, application of specific fungicides is recommended.

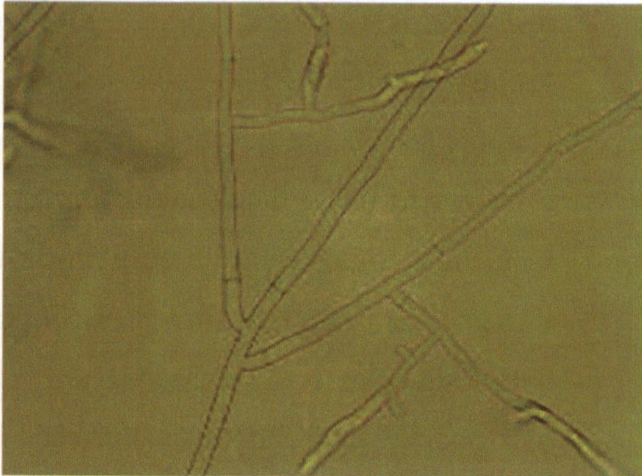

Figure 7-4. Symptoms and fungal isolates of rice sheath blight (*Pellicularia sasakii*) (National Crop Pest Management System)

5) Damping-off Disease

① Causal organisms is *Pythium debaryanum*, *Rhizoctonia solani*, *Fusarium* spp., fungi (Oomycetes and other groups)

② The pathogens overwinter as oospores in diseased tissues or soil and are transmitted through irrigation water or soil.

③ The disease occurs readily under conditions of low temperature with excessive moisture, poor aeration, alkaline soils, and excessive nitrogen fertilization.

④ It is prevalent in early-sown seedbeds and when there are large differences between day and night temperatures.

⑤ Symptoms include failure of seeds to germinate, or seedlings die before emerging from the soil.

⑥ Proper nursery management requires maintaining temperatures below 30°C during the day and above 10°C at night.

⑦ Control measures include the use of healthy seeds, sterilized nursery soil with a pH of about 4.5~5.5, and application of appropriate fungicides.

Figure 7-5. Symptoms of damping-off disease in rice (Pocheon News Today)

6) Bacterial Grain Rot

① Causal organism is *Burkholderia glumae*, bacterium (Pseudomonas, Proteobacteria).

② This pathogen is an aerobic, Gram-negative rod, and disease incidence is high at around 30°C.

③ Transmission occurs from infected to healthy seeds during soaking, and the disease develops about one week after heading under high temperature (30~35°C) and humid conditions.

④ Symptoms include drying or rotting of the whole or part of the panicle. Diseased grains become pale yellow to bluish-brown, and brown streaks appear on the kernels. In young seedlings, leaf sheaths may turn brown, leaves may become twisted or deformed, and plants may wither with yellowish-brown discoloration.

⑤ Control measures include the use of healthy seeds, seed disinfection, and avoidance of high temperature, high humidity, and excessive nitrogen fertilization during seedling cultivation.

⑥ In diseased fields, rice straw should be burned or removed after harvest.

Figure 7-6. Symptoms of bacterial grain rot (*Burkholderia glumae*) in rice (National Crop Pest Management System)

7) Bacterial Leaf Blight

① Causal organism is *Xanthomonas oryzae*, bacterium.

② In young rice plants, newly emerging leaves turn yellow, and unexpanded or expanded leaves show longitudinal yellow-green streaks.

③ Initially, the tips or margins of leaves develop pale yellow, grayish-white, or whitish wavy lesions, and in severe cases, the entire leaf dries up and dies.

④ The pathogen overwinters in weeds or rice stubble, and disease progression is promoted by typhoons, flooding, and excessive nitrogen fertilization.

⑤ The disease is favored by temperatures of about 26~30°C, and strong winds cause wounds on leaves that facilitate bacterial invasion, so outbreaks are frequent when typhoons and flooding occur together.

⑥ Control measures include removal of alternate hosts (e.g., Leersia, Zizania, *Echinochloa* spp.), avoidance of excessive nitrogen fertilization, and application of silicon and potassium fertilizers.

⑦ Burning weeds on levees helps eliminate overwintering vectors (e.g., leafhoppers), and cultivation of resistant varieties is recommended.

⑧ During the rainy season, when rice suffers from flooding injury, application of specific bactericides is required.

Figure 7-7. Symptoms of bacterial leaf blight (*Xanthomonas oryzae*) in rice (National Crop Pest Management System)

8) Black-Streaked Dwarf Disease

① Causal agent is Rice black streaked dwarf virus (RBSDV), virus.

② The disease is transmitted by the small brown planthopper (*Laodelphax striatellus*).

③ The vector overwinters in weeds, wheat fields, and Chinese milk vetch fields.

④ Symptoms include dark-green leaves with pronounced stunting. Brown or black streaks appear on the underside of leaves and on the culms.

⑤ Control measures include cultivation of resistant varieties and immediate removal of infected plants when detected.

⑥ Management also requires vector control and removal of weeds in paddy fields.

Figure 7-8. Symptoms of rice black-streaked dwarf disease (https://encrypted-tbn0. gstatic.com/images)

9) Dwarf Disease

① Causal agent is Rice dwarf virus (RDV), virus.

② The disease is transmitted by the green rice leafhopper (*Nephotettix cincticeps*) and the zigzag leafhopper (Recilia dorsalis).

③ The insect vectors overwinter in weeds, wheat fields, and Chinese milk vetch fields.

④ Symptoms include dark-green leaves with distinct white spots, severe stunting, excessive tillering, and plant height reduced to about half that of healthy rice before heading.

⑤ Control measures include cultivation of resistant varieties and removal of diseased plants.

⑥ Additional measures involve vector control, elimination of weeds on levees, and avoidance of excessive nitrogen fertilization.

10) Stripe Disease (Stripe Virus)

① Causal agent is Rice stripe virus (RSV), virus.

② The disease is transmitted transovarially by the small brown planthopper (*Laodelphax striatellus*), with an incubation period of about 7~14 days before symptoms appear.

Figure 7-9. Symptoms of rice dwarf disease

③ Disease incidence after transplanting, especially during the tillering stage, is largely determined by the density of viruliferous planthoppers.

④ Infected seedlings exhibit yellowing and rolling of inner leaves, with translucent yellowish stripes appearing. Panicles become deformed and grain setting is poor.

⑤ Control measures include management of the insect vector and removal of infected plants.

Figure 7-10. Symptoms of rice stripe disease caused by rice stripe virus (RSV) and transmitted by the small brown planthopper (*Laodelphax striatellus*) (National Crop Pest Management System)

2. Insect Pest Control

1) Insect pests occurring in rice

(1) Rice Planthoppers (Plant Hoppers)

A: Small Brown Planthopper (*Laodelphax striatellus*)

① The small brown planthopper is an overwintering insect pest that transmits Rice Stripe Disease (viral). Although recent overwintering densities are relatively low, when large populations migrate from China, serious damage is expected, requiring thorough preventive control.

② In areas where outbreaks are anticipated, it is advisable to plant resistant varieties such as Samkwang, Saenuri, Jopyeong, and Hwayoung, and to cover seedbeds with insect-proof nets near barley fields to block the inflow of planthoppers.

③ On the day of transplanting, granular insecticides should be applied, which can simultaneously control rice water weevils, rice leaf beetles, and leaf miners.

④ Damage symptoms include growth disorders caused by feeding on plant sap and transmission of viruses, especially Rice Stripe Disease.

⑤ In its life cycle, the insect overwinters as a 4th instar nymph and has 4~5 generations per year, occurring from April to September. Eggs are banana-shaped, white, about 0.6mm long, and the nymphs are light brown.

Figure 7-11. Nymphs and adults of the small brown planthopper (*Laodelphax striatellus*)

⑥ Adults are sexually dimorphic, with females having a pale yellow central body part while males are black, and body length ranges from 3 to 4 mm with females being slightly larger.

⑦ Control measures include spraying insecticides during the outbreak period and cultivating resistant varieties.

B: White-backed Planthopper (*Sogatella furcifera*)

① Damage symptoms: The insect feeds on plant sap, causing growth disorders and sometimes lodging.

② The life cycle shows that the insect migrates from China in mid to late June, as it cannot overwinter in Korea, and large populations are often observed after the passage of low-pressure systems. Accordingly, the number of occurrences varies depending on migration conditions, but generally about four generations occur annually.

③ Eggs are banana-shaped, white, about 0.9mm long, and nymphs are pale gray.

④ The dorsum bears dark brown spots, and adults are dark brown with a yellowish-white marking in the middle of the back, black lateral margins, and transparent wings.

⑤ Adult body size is about 3.5mm, with a characteristic white rhomboid pattern in the central dorsal region.

⑥ Control measures include cultivation of resistant varieties, reduced application of nitrogen fertilizer, and insecticide spraying during outbreak periods.

Figure 7-12. Adults, eggs, and damage symptoms of the white-backed planthopper (*Sogatella furcifera*) (National Crop Pest Management System)

C: Brown Planthopper (*Nilaparvata lugens*)

① Damage symptoms: This insect pest feeds on plant sap, causing growth disorders and often plant death.

② Life cycle: Since it cannot overwinter in Korea, it migrates in mid to late June. Although its reproductive capacity and mobility are relatively low, infestations can cause concentrated and severe damage. Generally, about four generations occur annually.

③ Eggs: They are banana-shaped, white, and approximately 0.6mm in length, and the nymphs are smaller than adults with shorter wings.

④ Adults: It is larger than the small brown planthopper, measuring 4.5~6mm in length, and its body color is brown with dark brown near the mouthparts and light brown wings.

⑤ Control: Cultivation of resistant varieties, reduced application of nitrogen fertilizer, and spraying of insecticides during outbreak periods.

⑥ Early control: Since migrating insects are critical sources of infestation, careful inspection of the lower stems of rice plants in areas where planthoppers have arrived is essential. If populations are high, registered insecticides should be applied promptly.

Figure 7-13. Adults, eggs, and damage symptoms of the brown planthopper (*Nilaparvata lugens*) (National Crop Pest Management System)

(2) Rice Water Weevil (*Lissorhoptrus oryzophilus*)

① The rice water weevil feeds on young rice leaves and roots, while the rice leaf beetle and leaf miners feed inside rice leaves or stems, causing recurring annual damage.

② For control, granular insecticides should be applied to nursery seedling boxes on the day of transplanting. If insecticides are not applied at the nursery stage, they should be sprayed 10~15 days after transplanting.

③ Because rice water weevils attack rice plants at an early stage (around transplanting), the resulting damage is particularly severe.

Figure 7-14. Adults, larvae, and damage symptoms of the rice water weevil (*Lissorhoptrus oryzophilus*) (National Crop Pest Management System)

(3) Rice Leaf Roller (*Cnaphalocrocis medinalis*)

① The rice leaf roller damages rice plants by rolling the leaves (also called "leaf folding") and feeding inside them.

② This pest migrates from China, overwinters in the pupal stage, and generally occurs two to four times per year.

③ The eggs are pale yellow and about 1.0mm in length. The larvae have light brown heads and yellow-green bodies, and when mature, they turn pale yellow.

④ For control, weeds on paddy levees should be removed during winter to kill overwintering insects, and insecticides should be applied during the outbreak period.

Figure 7-15. Adults, larvae, and damage symptoms of the rice leaf roller (*Cnaphalocrocis medinalis*) (National Crop Pest Management System)

(4) Green Rice Leafhopper (*Nephotettix cincticeps*)

① The green rice leafhopper feeds on the sap of rice plants, which causes growth disorders, plant death, and also transmits viral diseases such as rice dwarf virus.

② It has four generations per year. The eggs are banana-shaped, initially translucent but gradually turning yellowish-brown, and measure about 1.0mm in length. The nymphs are pale yellow after hatching but gradually change to dark brown.

③ The adults have green wings with brown or black tips. Males are about 4~5mm long, while females are about 6 mm.

④ During the outbreak period, insecticides should be applied.

Figure 7-16. Adults(black-tipped wings are males), nymphs, and damage symptoms of the green rice leafhopper (*Nephotettix cincticeps*) (National Crop Pest Management System)

(5) Rice Stem Borer (*Chilo suppressalis*)

① When the larvae bore into rice plants, the leaves turn yellowish-brown, wither, and eventually die.

② The insect has two generations per year. It overwinters as mature larvae in rice straw. The first generation hatches from pupae in April, appears in May, and reaches its peak in June, spreading by wind. The second generation emerges in July and reaches its peak in August.

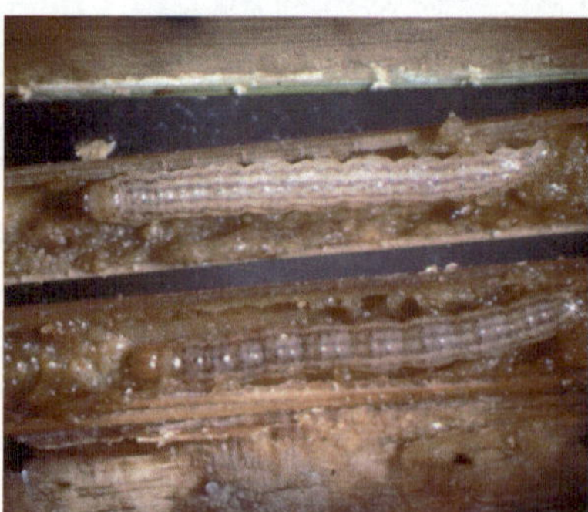

Figure 7-17. Adults, larvae, and damage symptoms of the rice stem borer (*Chilo suppressalis*) (National Crop Pest Management System)

③ About 10~18 days after the peak of egg hatching, insecticides should be applied to stems and leaves for control.

(6) Storage Pests

A: Weevils

① Weevils are insect pests that damage stored grains and belong to the family Curculionidae of the order Coleoptera. The rice weevil (*Sitophilus oryzae*, formerly *Calandra oryzae*) is distributed worldwide.

② They possess hardened forewings (elytra), and females lay 300~500 eggs by boring holes into rice kernels.

③ As chewing insects, both larvae and adults cause damage, particularly in dark conditions.

④ They overwinter as larvae or adults, and their activity increases when temperatures rise to 15~16°C. From mid- to late May, they oviposit inside grain kernels, and the hatched larvae feed and grow inside.

⑤ Three to four generations occur annually, with the optimum temperature for reproduction being 28~29°C.

⑥ For control, grains should be thoroughly dried and stored at a moisture content of 12% or less.

⑦ Storage facilities should be cleaned, ventilated, and kept at low temperatures.

⑧ Adults can be trapped and killed, or parasitoids (natural enemies) that attack larvae can be utilized.

⑨ Storage structures can be sealed and fumigated with chloropicrin or carbon disulfide.

Figure 7-18. Weevil infestation in wheat and adults of the rice weevil (*Sitophilus oryzae*)

B: Indian Meal Moth (*Plodia interpunctella*)

① This insect belongs to the family Pyralidae (order Lepidoptera). The larvae are about 15mm long and have reddish-brown heads.

② The adult lifespan is less than two weeks, but oviposition begins the day after mating, with about 150 eggs laid per female.

③ Adults prefer dark environments.

④ Commonly called the rice moth (*Plodia interpunctella*), it is the most destructive storage pest of rice. The larvae damage grains and reduce their quality.

⑤ It has three to four generations per year. The optimum temperature for reproduction is 28~33°C, and higher humidity increases infestation severity.

⑥ The insect overwinters in the larval stage, spinning a thin cocoon in which it remains. Individuals that pupate in late autumn overwinter as pupae.

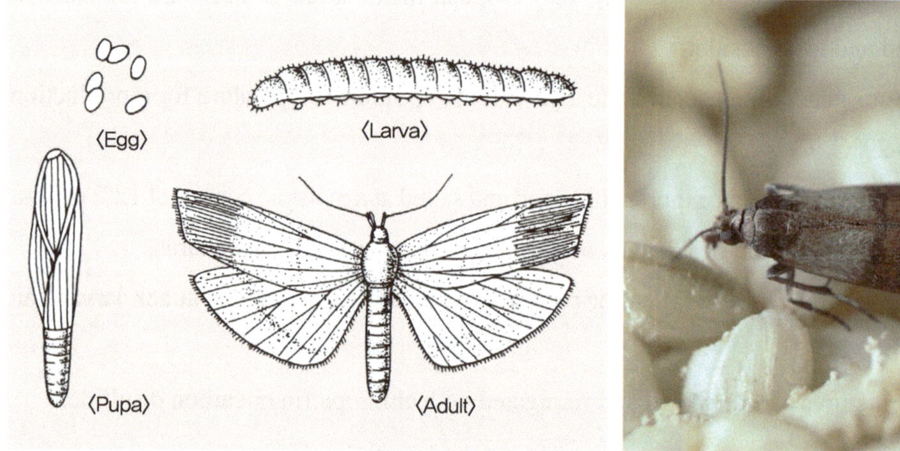

Figure 7-19. Indian meal moth (*Plodia interpunctella*)

⑦ For control, rice should be dried to a moisture content of about 12% before storage.

⑧ Biological control using parasitoid wasps or chemical insecticides can be applied for suppression.

C: Flour Beetle (*Tribolium* spp.)

① The flour beetle occurs several times throughout the year, is omnivorous, and has a strong reproductive capacity.

② When it bores into rice, it leaves behind a large amount of powdery residue, which provides favorable conditions for other pests.

③ Rice heavily infested by this insect develops an unpleasant odor.

④ Once infestation occurs, significant damage results, including quantitative loss from feeding, contamination and odor caused by excreta, and secondary infections and infestations.

⑤ Most pests cannot reproduce well when the grain moisture content is below 12% (relative humidity under 55%), but the flour beetle breeds vigorously when moisture content exceeds 14% (relative humidity around 75%).

Figure 7-20. Flour beetle (*Tribolium castaneum* or *Tribolium confusum*)

3. Weed Control

1) Definition and damage of weeds

① A weed is defined as an unwanted plant, one that bears no useful product or grows outside of its proper place.

② In contrast, a crop is a plant cultivated by humans for its useful value, with certain plant parts artificially developed to an almost abnormal extent for human use.

③ Weeds are plants that interfere with crop growth, such as barnyardgrass (*Echinochloa crus-galli*) in rice paddies or crabgrass (*Digitaria* spp.) in barley fields.

④ The damage caused by weeds has been increasing, and herbicides account for the highest sales among pesticide formulations, mainly due to rising labor costs and the need to reduce manual weeding.

⑤ Weeds compete with crops for light, water, and nutrients, degrade crop quality, and serve as habitats for pests and diseases, ultimately hindering crop growth and reducing yields.

⑥ In other words, weeds are unwanted plants or plants growing out of place, and their removal is beneficial for crop cultivation. They are herbaceous plants that contradict human intentions for land use and occur in farmland or surrounding environments.

⑦ By directly or indirectly damaging crops, weeds reduce productivity and oppose human management of land, thereby lowering both land productivity and labor productivity.

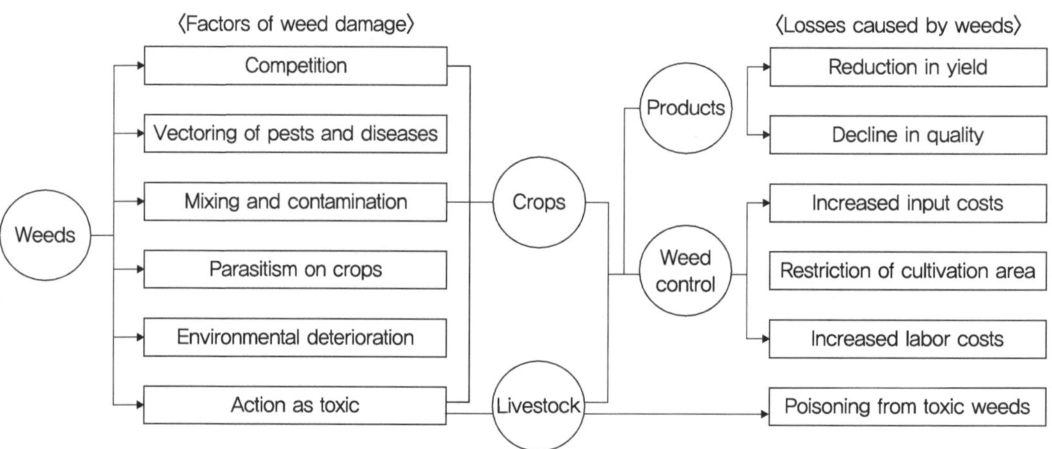

Figure 7-21. Factors and losses caused by weeds

⑧ Weeds arise naturally in cultivated fields and cause both yield reduction and quality deterioration of crops, either directly or indirectly.

⑨ Weeds are highly diverse, and therefore classified physiologically, ecologically, and morphologically into grass weeds (Gramineae), broadleaf weeds, and sedges (Cyperaceae). According to life cycle, they are further divided into annual, biennial, and perennial weeds.

⑩ The history of agriculture is often described as the history of weed control, as weeds have always been troublesome and burdensome in farming. Many people who return to farming or begin rural life describe their struggle as a "war against weeds," highlighting the severity of the problem.

2) Classification of Weeds

(1) Barnyardgrass (*Echinochloa oryzicola*)

① This grass weed (family Poaceae) is a summer annual that occurs widely in paddy fields across Korea.

② The culms grow 50~110cm tall and form dense clumps resembling rice plants.

③ The leaf blades are light green, and the ligules are absent. The leaf sheath margins are smooth and glabrous, while the surface is rough to the touch. The inflorescence is linear, measuring 10~20cm in length and 8~12mm in width.

④ The flowers bloom from august to october and are borne on greenish panicles 8~15cm long.

⑤ The spikelets are long-ovate, 4~5.5mm in length, each bearing two florets.

⑥ The spikelet apex is pointed and glossy, and the awns are either absent or very short when present.

Figure 7-22. Inflorescence, seeds, and growth habit of barnyardgrass (*Echinochloa oryzicola*) (Encyclopedia of Korean National Culture)

(2) Long-awned Barnyardgrass (*Echinochloa crus-galli* var. *echinata*)

① This grass weed (family Poaceae) is a summer annual that occurs in paddy fields throughout Korea.

② The culms grow 80~150cm tall, cluster together, and spread somewhat obliquely from the base. The leaf blades are 30~50cm long and 10~20mm wide, and ligules are absent.

③ The leaf sheaths are glabrous, with the basal portion tinged reddish-purple.

④ The flowers bloom from July to September and are borne on panicles 10~25cm long. The spikelets are reddish-purple, ovoid, 3~4mm long, with bristle-like transparent hairs on the surface, and contain two florets.

⑤ The spikelet tips bear awns 2~4cm in length.

Figure 7-23. Inflorescence, seeds, and growth habit of barnyardgrass (*Echinochloa crus-galli* var. echinata) (Illustrated Guide to Korean Plant Ecology)

(3) Common Barnyardgrass (*Echinochloa crus-galli*)

① This grass weed (family Poaceae) is a summer annual that propagates by seed.

② It grows abundantly in rice paddies, ditches, upland fields, field margins, pastures, roadsides, orchards, and fallow land. As a weed well adapted to wetlands, it thrives particularly in low-lying moist areas, including paddy fields.

③ It grows not only under submerged conditions but also in dry habitats such as orchards, uplands, or roadsides. However, plants in dry areas are shorter in height and produce fewer seeds, panicles, and tillers.

④ For seed germination in dry sites, oxygen is required, but in wet or shallow-water conditions, such as in paddy fields, it can germinate even in the absence of oxygen. It grows best in sunny, moist, and fertile soils, especially sandy soils rich in nitrogen. Severe infestations in soybean fields or rice paddies cause significant crop damage.

Figure 7-24. Inflorescence, seeds, and growth habit of common barnyardgrass (*Echinochloa crus-galli*) (Illustrated Guide to Korean Plant Ecology)

3) Monochoria (*Monochoria korsakowii*)

① This annual summer weed (family Pontederiaceae) occurs nationwide in Korea, with particularly heavy infestations in paddy fields along the west coast.

② The stems grow 20~40cm tall. The leaf blades are glossy and cordate, measuring 5~13cm in both length and width, with acutely pointed apices and bases deeply notched in a heart shape.

③ Bluish-purple flowers bloom from September to October in panicles 5~15cm long borne at the tips of main stems. Leaf-like bracts are present at the base of the inflorescence.

④ Each flower is 2.5~3cm in diameter, with six elliptic petals. The fruit is about 1cm long.

Figure 7-25. Growth and inflorescence of water hyacinth (*Eichhornia crassipes*) (Illustrated Guide to Korean Plant Ecology)

4) Indian Jointvetch (*Aeschynomene indica*)

① Indian jointvetch, a member of the family Fabaceae, is a summer annual weed that occurs widely in paddy fields throughout Korea.

② The stems are erect, 50~200cm tall, and hollow in the upper portions.

③ The leaves are alternate, each consisting of 10~20 pairs of small leaflets arranged in a pinnate pattern. The leaflets are 1~1.5cm long and 2~3.5mm wide, yellowish on the surface with a whitish bloom on the underside.

④ The stipules are ovate or lanceolate, 7~12mm long, with pointed tips.

⑤ The pale yellow, butterfly-shaped flowers appear from July to August, usually 2~3 at the ends of peduncles.

⑥ The fruits are flat, linear pods with 6~8 joints, and when mature, the joints separate individually.

Figure 7-26. Leaves, flowers, fruits, and growth habit of Indian jointvetch (*Aeschynomene indica*) (Doosan Encyclopedia)

5) Beggar-ticks (*Bidens frondosa*)

① *Bidens frondosa* (family Asteraceae) is a summer annual alien weed that infests paddy fields throughout Korea.

② The stems are erect, 50~200cm tall, quadrangular, glabrous, and dark purplish in color.

③ The leaves are opposite and pinnately compound with 3~5 leaflets. The leaflets are lanceolate, 3~13cm long, with long acuminate tips, bases that narrow abruptly, serrated margins, and distinct petiolules.

④ The yellow flowers bloom from June to October and are borne at the ends of branches and stems.

⑤ The achenes are flattened, 6~10mm long, with two barbed awns at the apex.

Figure 7-27. Leaves, flowers, seeds, and growth habit of beggar-ticks (*Bidens frondosa*) (Source: Illustrated Guide to Korean Plant Ecology)

6) Creeping Primrose-willow (*Ludwigia prostrata*)

① This summer annual weed (family Onagraceae) infests paddy fields throughout Korea.

② The stems are 30~100cm tall, reddish in color, and marked with longitudinal ridges.

Figure 7-28. Early growth, flowers, and growth habit of creeping primrose-willow (*Ludwigia prostrata*) (Source: Doosan Encyclopedia)

③ The leaves are alternate, lanceolate, 3~12cm long and 1~3cm wide, tapering at both ends, with entire margins, and often turning reddish in autumn. The petioles are 5~15mm long.

④ The yellow flowers bloom in September, about 1cm in diameter, borne singly in the leaf axils, and lack pedicels.

⑤ The fruits are elongated-ovoid capsules, 1.5~3cm long and 1.5~2mm wide.

7) Monochoria (*Monochoria vaginalis* var. *plantaginea*)

① This summer annual weed (family Pontederiaceae) grows in shallow water of paddy fields, fallow paddies, ditches, wetlands, and ponds.

② It grows well regardless of soil type and has strong reproductive capacity. In rice fields, if not controlled, it can spread throughout the paddy and seriously reduce yields. Because it is highly adapted to fields where chemical fertilizers are continuously applied or where soil fertility is high, it occurs in large numbers under such conditions. Mature fruits easily detach under slight water movement or mechanical impact.

③ When fruits fall into water or wet areas, they absorb moisture and swell, causing the pericarp to split backward and release the seeds. The released seeds are dispersed by water.

④ Resistant ecotypes to sulfonylurea herbicides have emerged, posing a serious problem.

Figure 7-29. Growth habit of *Monochoria vaginalis* var. plantaginea (Source: Illustrated Guide to Korean Plant Ecology)

8) Arrowhead (*Sagittaria trifolia*)

① This perennial weed (family Alismataceae) grows in paddy fields, wetlands, or shallow water. It reproduces both by seeds and by tubers formed at the tips of underground rhizomes.

② During growth, the leaf shape changes three times. In the early stage, seedlings emerging from the scape produce 5~10 narrow, linear leaves, followed by 1~2 spatulate leaves with broader blades. Subsequently, sagittate (arrowhead-shaped) leaves develop, gradually increasing in size to form the typical morphology of Sagittaria.

③ It is one of the major dominant weeds in paddy fields throughout Korea. Unlike some other species, it does not propagate vegetatively by producing new tillers during the growing season.

④ This nitrogen-loving weed competes with rice for nutrients. Because it grows to a similar height as rice, it shades rice plants and interferes with photosynthesis, thereby reducing rice yields.

Figure 7-30. Growth habit and flowers of arrowhead (*Sagittaria trifolia*).

9) Pondweed (*Potamogeton distinctus*)

① This perennial weed (family Potamogetonaceae) grows in paddy fields, ditches, and shallow wetlands, reproducing mainly through tuber-like propagules.

② The rhizomes spread laterally through the soil beneath the water, often forming large colonies.

③ From late summer to early autumn, the tips of the rhizomes swell and curve downward, producing numerous yellowish, banana-shaped tubers. Because rhizomes and submerged stems regenerate easily by producing new roots from the nodes after being cut, manual removal of pondweed from paddy fields is very difficult.

④ Cut tubers sprout faster and show more vigorous early growth than intact tubers, enabling them to form large colonies readily. This characteristic has become a major factor in the spread of pondweed, especially as mechanized rice farming increases the frequency of rhizome fragmentation.

Figure 7-31. Appearance of the perennial broadleaf weed pondweed (*Potamogeton distinctus*) (Source: Doosan Encyclopedia)

4. Meteorological Disasters

1) Temperature Stress

Temperature stress can generally be divided into low-temperature injury and high-temperature injury, which interact in a compensatory relationship with light intensity. Such growth interactions also appear under drought and heat stress, meaning that drought and heat damage occur more readily together than separately, and the damage is far more severe. Low-temperature stress includes chilling injury, cold-weather damage (caused by exposure to temperatures lower than the optimum without freezing), and freezing injury (caused by actual cell freezing). When temperatures exceed the optimum for growth, high-temperature injury occurs, and at even higher temperatures, plants may die.

(1) Cold-weather damage

Cold-weather damage is a type of low-temperature injury occurring at temperatures above 0 °C without freezing, mainly affecting summer crops. Rice, as a subtropical crop, suffers cold damage when it is exposed to low temperatures during sensitive growth stages in temperate regions.

A: Definition of cold-weather damage

① When leaves are exposed to low temperatures, photosynthesis and translocation of assimilates are reduced, respiration and protein synthesis decline, protein degradation is promoted, and physical changes in lipids reduce membrane fluidity. As a result, membrane proteins and enzymes lose function. Plants with low cold tolerance tend to have a higher proportion of saturated fatty acids, which crystallize at low temperature, further reducing membrane fluidity.

② When air temperatures are high but soil temperatures are low during the nursery period, high water viscosity reduces water absorption while transpiration increases, resulting in water imbalance. Rice leaves may wilt under these conditions, and in severe cases, plants die.

③ Cold injury can be reduced by flooding fields to submerge leaves, which raises water temperature, or by applying abscisic acid (ABA) to suppress transpiration.

B: Types of cold-weather damage in rice

① Cold injury in rice is classified into delayed type, sterility type, disease type, and mixed type, the latter being a combination of the others.

② Delayed type cold injury occurs during vegetative growth stages such as seedling, transplanting, and tillering, delaying heading and reducing grain filling, ultimately decreasing yield.

③ Sterility type cold injury arises from abnormal reproductive organ development. Low temperatures during germination or meiosis result in poor germination and leaf discoloration. At meiosis or panicle initiation, the tapetum layer enlarges and pollen becomes sterile, preventing anther dehiscence and pollination. At flowering, low temperatures prevent spikelet opening, causing sterility, while low temperatures during grain filling reduce grain weight and quality.

④ Disease type cold injury occurs because low temperatures inhibit photosynthesis and protein synthesis, leading to accumulation of soluble amino acids and amides. The rice blast fungus can utilize amino acids directly, and under such conditions, blast disease outbreaks occur.

C: Cold tolerance

① Cold injury involves interconnected biochemical processes (photosynthesis, respiration, protein and lipid metabolism) at temperatures below the optimum. Its physiological basis is complex, and the mechanisms of cold tolerance remain unclear.

② Cold tolerance differs by cultivar. *Japonica* rice is more tolerant than *Indica* and Tongil-type varieties. In Korea, during the 1980s, extensive cultivation of Tongil-type rice led to yield reductions of up to 50% due to cold injury, increasing awareness of meteorological disasters and stress physiology.

③ Crops with strong cold tolerance, such as cauliflower, turnip, and pea, have higher proportions of unsaturated fatty acids than crops with weak tolerance, such as kidney bean, sweet potato, and maize. When crops are exposed to low temperatures, the proportion of unsaturated fatty acids increases during hardening, enhancing cold tolerance.

④ Proline content increases under drought but decreases markedly in rice anthers under low temperatures.

D: Countermeasures against cold injury

① Selection of cold-tolerant cultivars: For example, *Japonica* varieties show greater tolerance than Tongil-type varieties, which suffered severe damage in the 1980s when widely cultivated during rice shortages.

② Improvement of field conditions: Establish windbreaks, improve poorly drained or leaky paddies, and build soil fertility.

③ Improvement of cultivation methods: Adopt protected seedling raising, early or short-duration cultivation to advance maturity, and use dense planting to promote vigorous growth.

④ Temperature control: Maintain higher night temperatures in cool periods by deep flooding, raising irrigation water temperature using reservoirs, or applying evaporation retardants.

2) Water Stress

(1) Drought stress

When soil or air is dry, plant water content decreases, leading to growth inhibition and, in severe cases, wilting and death. This damage due to water shortage is termed drought stress, and the ability of crops to withstand it is called drought tolerance. Symptoms of drought stress include:

① Reduced cell growth, suppressing overall crop growth.

② Reduced biosynthesis and enhanced hydrolysis.

③ Suppressed transpiration and photosynthesis.

④ Inhibited cell division and accelerated senescence.

⑤ Accelerated translocation of water and nutrients, leading to earlier fruit ripening.

⑥ Increased proline and protein content, which serve as indicators of drought stress.

⑦ Increased translocation of assimilates and higher respiration.

⑧ When drought stress occurs during vegetative growth, leaves become smaller, and during maturity the leaf area index (LAI) decreases, reducing light interception and photosynthesis.

⑨ Chlorophyll synthesis is inhibited under severe water deficiency.

⑩ Most enzymes lose activity under water stress (e.g., nitrate reductase), but the activity of some hydrolytic enzymes, such as amylase, may actually increase.

Table 7-3. Effects of water stress on crop metabolic processes

Metabolic process or paramete	Tissue water potential (bars)				Remarks
	-5	-10	-15	-20	
Cell growth decreases					Ripid growing tissue
Cell wall synthesis decreases					
Protein synthesis decreases					Chlorosis of leaf
Chlorophyll synthesis decreases					···
Nitrate reductase level decreases					···
ABA synthesis					···
Stomatal closure					Depends on the species
CO_2 fixation decreases					
Respiration decreases					···
Xylem transport decreases					···
Proline accumulation					···
Sugar concentration increases					···

Source: Hsiao et al. 1976.

(2) Mechanisms of Drought Tolerance

① When water becomes limited, crops undergo physical and biochemical changes that allow them to adapt to dry conditions. Even within the same species, seeds or dormant cells can endure desiccation better than actively growing cells, which are more vulnerable to drought stress.

② Depending on the severity of water shortage, crops adopt different strategies. Under mild drought, they suppress leaf expansion or close stomata to reduce transpiration while extending roots deeper to reach available water.

③ As soils dry further and approach the permanent wilting point (around –15 bars), water absorption nearly ceases, resulting in leaf wilting.

④ During midday, when transpiration is high, root shrinkage may occur, separating root hairs from soil particles. Root hairs are stressed, and suberin accumulates in the outer cortex, reducing water permeability and making water absorption more difficult.

⑤ Under severe drought (soil water potential between –20 and –10 bars), water tension in both plants and soil increases, and leaves suffer water deficit damage. Continued reduction in plant water content leads to plasmolysis, structural collapse of cells, and eventual death. Cultivar differences exist, and for example, Namphung rice exhibits stronger drought tolerance than Chucheong rice.

(3) Types and Responses of Drought Tolerance

① Based on water requirements during growth, plants are classified as hydrophytes, mesophytes, or xerophytes.

② Without water limitation, xerophytes transpire more than mesophytes, but under wilting conditions, xerophytes transpire less. Thus, drought tolerance depends on the plant's ability to restrict transpiration under water deficit.

③ Drought tolerance is broadly categorized into structural (constitutional drought tolerance) and protoplasmic (true physiological drought tolerance).

④ Structural drought tolerance involves delaying or avoiding water loss. Crops with small leaf areas, low water requirements, or low transpiration conserve soil water early and utilize it later during summer droughts. Such traits are typical of succulents.

⑤ Crops with well-developed deep root systems maintain water absorption despite high transpiration rates, thereby exhibiting drought resistance.

⑥ Protoplasmic drought tolerance refers to true drought tolerance at the cellular level, where reduced water content in the protoplasm does not cause serious physiological injury.

⑦ Drought damage manifests not only in large-scale yield reduction but also in diverse physiological impairments, the most common being reduced or arrested cell growth, which leads to secondary injuries.

A: Stomatal closure

① When relative humidity is low and wind speed is high, guard cells lose water faster than they can be replenished from neighboring cells, causing stomatal closure. This is passive closure due to guard cell dehydration in dry air.

② Under water deficit, ABA (abscisic acid) is synthesized and reduces membrane selectivity, causing solutes in guard cells to diffuse outward. As guard cell water potential increases, water exits, turgor is lost, and stomata close. This is active closure.

③ In normal leaves, mesophyll cells also produce ABA, which accumulates in chloroplasts. When water becomes limiting, ABA migrates into the cell wall and, via the transpiration stream, reaches guard cells, inducing closure. Under drought, ABA levels may increase up to 40-fold. Initially, ABA translocated from chloroplasts causes closure, and later newly synthesized ABA maintains it.

④ Continued drought reduces leaf water, lowers chloroplast pH, increases ABA synthesis, and drives ABA into the cytoplasm, then into the cell wall, from where it moves with transpiration to guard cells, maintaining closure.

⑤ ABA synthesized in roots also promotes stomatal closure. Even when only part of the root system experiences drought, ABA produced there is transported upward via the transpiration stream, inducing stomatal closure in leaves.

B: Reduction of leaf area

① Water shortage lowers leaf turgor, suppressing cell expansion and reducing leaf area. With stomata closed, transpiration is minimized, but even after rehydration, thickened cell walls reduce elasticity and slow subsequent growth.

② Compared with well-watered plants, cereals show leaf rolling, while dicots display drooping leaves, reducing light interception. Despite sunlight, stomata remain closed, photosynthesis is inhibited, and leaf temperature rises, increasing respiration.

③ Leaf rolling and reduced leaf expansion decrease light interception and alleviate heat injury, thereby mitigating drought stress.

C: Increased relative root growth

① Photosynthesis is less sensitive to turgor than leaf expansion, so under water deficit, leaves may not elongate but photosynthesis continues to some extent.

② Carbohydrates that would normally support leaf expansion are redirected to roots, enhancing root elongation and enabling water uptake from deeper layers. Root shrinkage in dry surface soil promotes downward extension, improving drought adaptation.

③ n cacti, roots are shallow and widely spread near the surface, allowing rapid water uptake after even small rainfall events.

D: Wax accumulation and reduced cuticular transpiration

① Under drought, wax accumulates on leaf surfaces, reducing cuticular transpiration.

② In CAM plants, leaves are reduced to spines or stems with small surface area, deeply sunken stomata, thick cuticles, and well-developed water-storage tissues, all of which minimize water loss and enhance drought survival.

③ Suberin accumulates in roots, improving water conservation under drought conditions.

E: Agronomic practices for drought avoidance

① In rice, the panicle initiation stage is particularly sensitive to drought, so management practices should aim to avoid water deficit during this period. In upland crops, uneven rainfall distribution often causes temporary water shortages.

② Determinate soybean cultivars suffer yield loss if drought occurs during flowering, but in Korea, drought periods are short and not annually recurrent, so determinate types are mainly cultivated.

③ Indeterminate cultivars can escape drought damage by flowering and setting pods after rains return, but in Korea they are less suitable because early autumn frost before seed maturity reduces yield.

F: Accumulation of organic solutes in the protoplasm

① When water is deficient, cell size decreases, cell sap concentration increases, and solute osmotic potential decreases.

② Due to structural and compositional changes in enzymes, the activity of degradative enzymes increases, leading to higher absolute concentrations of sugars, organic acids, and inorganic salts such as potassium. As a result, osmotic potential decreases further, enhancing water absorption from the soil.

③ In the cytoplasm, ions are stored mainly in vacuoles, while osmotic organic solutes such as proline, sorbitol, glycine, and betaine accumulate in the cytoplasm.

④ Since proline synthesis increases under drought, it can be used as an indicator in breeding programs for drought tolerance. These osmotic solutes help lower the cytoplasmic water potential through osmotic adjustment with vacuolar ions, thereby facilitating water uptake via roots and enhancing drought tolerance.

(4) Countermeasures Against Drought Injury

A: Rapid irrigation

When crops suffer from drought, irrigation (surface, sprinkler, or subsurface) should be applied promptly. Because soil water potential decreases exponentially and recovery is difficult once the permanent wilting point is passed, irrigation should be carried out quickly considering farm management efficiency.

B: Hardening treatment of crops

Hardening makes crops more resistant by reducing cell and stomatal size, increasing stomatal density, and raising cell sap concentration. This treatment should be applied during seedling raising to develop strong root systems and enhance drought tolerance.

C: Improving soil water retention and reducing evaporation

Forming soil aggregates and mulching prevents evaporation. Intermediate tillage or dry farming may initially increase surface evaporation but ultimately reduces continuous capillary water loss. In paddy fields, evaporation retardants (e.g., oil films) may be applied to reduce water loss.

D: Selection of drought-tolerant cultivars

Drought-tolerant cultivars maintain relatively stable photosynthesis and respiration under low protoplasmic water content, delay protein and sugar loss, and therefore should be selected for cultivation.

(5) Flood Injury

A: Types: Submergence and complete flooding

① Flood injury includes partial submergence (submergence injury) and complete inundation (flooding injury).

② Combined submergence and flooding often occur with storms and can also be considered storm damage.

③ Typhoons and floods may lodge rice plants or cause mechanical injury, predisposing them to pests and diseases.

B: Occurrence and damage

① During submergence, photosynthesis and respiration are impaired due to insufficient oxygen, forcing plants into anaerobic respiration. Respiratory substrates are depleted, plant growth is suppressed, and susceptibility to pests and diseases increases, leading to yield loss.

② Damage varies depending on water quality and plant condition, often manifesting in complex forms.

③ The severity of injury depends on factors such as flooding duration, water temperature, water quality, flow velocity, growth stage, plant health, and cultivar.

④ Generally, complete flooding causes more damage than partial submergence, turbid water is worse than clear water, stagnant water is worse than flowing water, and higher water temperatures cause greater injury than cooler conditions. When these factors act together, damage is further aggravated.

⑤ "Green blight" occurs under turbid, stagnant, and warm water, where oxygen deficiency is severe and only carbohydrates are consumed while proteins remain undecomposed.

⑥ "Red blight" occurs under temporary flooding with clear or flowing water and lower temperatures. Here, partial photosynthesis continues while carbohydrates are slowly consumed, followed by protein depletion, causing leaves to turn reddish-brown.

C: Flooding damage by growth stage

① Immediately after transplanting, rice is least affected by flooding.

② At the tillering stage, tiller production is delayed or suppressed, reducing panicle number.

Just 3`4 days of flooding may reduce yields by 20–30%.

③ At the panicle initiation stage, branch and spikelet differentiation are reduced. If leaf tips remain above water, 3~4 days of flooding reduce yield by 20–30%, but complete submergence in turbid water can reduce yield by up to 50%.

④ The meiotic stage is most vulnerable: 3~4 days of flooding in turbid water can cause 50% yield loss. Under the same conditions, losses are about 50% at heading and 30% at grain filling.

⑤ During grain filling, frequent rainfall or lodging can induce pre-harvest sprouting, significantly reducing grain quality.

D: Countermeasures against flood injury

① Drain submerged paddy fields promptly and wash off soil and debris from leaves before they dry, in order to restore photosynthesis.

② Select cultivars with resistance to flooding, diseases, insects, and lodging. Mixed cultivation of 2~3 cultivars with different heading dates and lodging resistance can reduce risk.

③ After drainage, supply oxygen-rich water to restore root activity, and apply pesticides to control blast, bacterial leaf blight, and other diseases.

④ For lodged rice, tie 4–6 clumps together to raise them, increase application of potassium and silicate fertilizers, reduce nitrogen fertilization, and maintain balanced nutrient management.

(6) Hail Injury

A: Occurrence and damage

① In Korea, hail occurs locally in May~June or September~October, causing crop damage.

② The larger the hailstones, the higher the cloud humidity, and strong updrafts and longer travel distance through clouds also increase hail size. Large hailstones tend to fall in the afternoon when air temperatures are higher.

B: Countermeasures

When hail damages leaves, apply fungicides to prevent blast and provide foliar fertilization for nutrition. If grains have partially shattered, consider early harvesting.

3) Wind Stress

(1) Wind Injury

A: Occurrence and damage

① Wind injury causes mechanical damage to stems and leaves, lodging, shattering, and discolored grains.

② Damage is most severe from heading through maturity.

③ Wind reduces photosynthesis, increases disease incidence, and when lodging occurs, grain filling is poor, yields decline, grain quality decreases, and combine harvesting becomes difficult.

B: Foehn phenomenon (abnormally dry wind)

① When typhoon air ascends the Taebaek Mountains, releases rain, and descends the leeward side, it warms, becomes extremely dry, and transforms into a foehn wind.

② As moist maritime air loses moisture as rainfall, then warms over land, relative humidity drops sharply, sometimes causing "white head" in rice.

③ Typically, foehn damage occurs when night temperature is above 25°C, relative humidity below 65%, and wind speed 4~8m/s. If this occurs within 3~4 days after heading, damage is most severe.

C: White head phenomenon

① Abnormally dry winds around heading cause sterility due to failure of fertilization or arrested ovary development, producing sterile panicles ("white heads"), leading to severe yield loss.

② White head also occurs during severe rice blast infection, stink bug infestation, and second-generation rice stem borer damage.

D: Discolored grains

① Strong winds during early grain filling cause brown discoloration from mechanical friction and secondary infection.

② If typhoons occur 1~2 days after heading, discolored grain incidence increases sharply.

③ A higher proportion of discolored grains reduces yield, lowers grain filling and head rice ratio, and degrades quality.

E: Lodging

① Lodging results from bending or breaking of stems and involves diverse, complex factors.

② Lodging occurs together with wind injury, especially after stem elongation is complete, mainly after heading.

③ It is more frequent with excessive nitrogen fertilization, late transplanting, and dense planting.

④ Tall plants with weak stems and heavy panicles are more prone to lodging, particularly under strong wind and rain.

F: Countermeasures against wind injury

① Select lodging-resistant cultivars, and mix 2~3 cultivars differing in heading time and lodging resistance.

② Avoid dense planting and excessive nitrogen, particularly limiting nitrogen application during internode elongation.

③ Apply intermittent irrigation and midseason drainage to strengthen roots and stems and reduce nitrogen uptake.

④ When a typhoon is expected, deep irrigation reduces white head and lodging, and lodging-prevention chemicals may be applied if necessary.

⑤ If lodging occurs, establish drainage canals to remove water quickly, and if plants are completely lodged, tie 4~6 clumps together to raise them.

(2) Salt Wind Damage

A: Occurrence and damage

① When strong winds blow inland from the sea, rice can suffer from salt wind damage due to winds carrying saline particles.

② Strong winds without rainfall cause not only mechanical wind injury but also additional salt damage, as seawater salts are carried in the air, leading to plant death.

③ Damage from salt winds can occur up to 2km inland, with severity increasing in fields closer to the coast.

④ Symptoms include leaf browning or death, sterile panicles ("white heads"), and discolored grains.

⑤ When winds travel over cold seawater, they become cold salt winds, which can combine salt injury with cold injury, and this phenomenon is sometimes observed along Korea's east coast.

B: Countermeasures

If salt wind damage occurs, spraying fresh water to wash off salts can prevent reduced grain filling and help minimize yield losses.

Rice Crop Science

Chapter 8. **Special Cultivation and Eco-Friendly Cultivation**

Chapter 8.
Special Cultivation and Eco-Friendly Cultivation

With changing times and diverse farming technologies, new cultivation methods have continuously developed. Among them, special cultivation and eco-friendly cultivation for safe food production are emerging as current research directions. Direct seeding cultivation is seen as a method that allows labor-saving practices and expansion of cultivation areas through efficient use of agricultural machinery. Meanwhile, eco-friendly cultivation, which minimizes the use of agricultural inputs while maintaining yield and safety, has become an unavoidable trend. These cultivation systems have both advantages and disadvantages, and it is necessary to develop techniques that enhance their strengths while compensating for their weaknesses. To achieve this, an understanding of basic cultural practices is essential, which will be examined in this chapter.

1. Cultural Practice

1) Early and Late-Early Season Culture

① In early and late-early season culture, seedlings are raised under low-temperature conditions, so attention must be paid to protecting seedbeds from cold, and cultivars with good low-temperature germination ability should be selected.

② Since the growing period is extended, this practice is generally effective in fertile paddy fields, and fertilizer application needs to be increased by 20~30%.

③ Because vegetative growth is abundant, plants are prone to excessive tillering, pest and disease incidence increases, lodging is more likely, and grain quality may be somewhat reduced.

(1) Early Season Culture

① This method involves sowing and raising seedlings of early-maturing varieties (generally temperature-sensitive cultivars) as early as possible, transplanting them early, and harvesting rice at an early stage.

② It is practiced in high-altitude, cool northern regions or southern plains to facilitate double cropping of rice with upland crops.

③ The grain-filling period in early season culture is about 30~35 days, which is relatively short. In plain regions, however, grain filling occurs during hot weather, often resulting in reduced eating quality.

Table 8-1. Classification of rice cultural practices by cropping season

(△: Sowing, O: Transplanting, ●: Heading, □: Harvesting)

Cultural practice	Apr.	May	Jun	Jul.	Aug.	Oct.	Sep.
Normal-season culture							
Early-season culture							
Early planting culture							
Late planting culture							
Late-season culture							

(2) Early Planting Culture

① Early planting culture refers to transplanting seedlings of medium- to late-maturing rice varieties at an earlier date in cold regions.

② This practice extends the overall growing period and aims at higher yields, while slightly advancing the heading stage, thereby reducing the risk of cold injury during the late growth phase in cold regions.

③ It also enables farmers to avoid damage from typhoons or certain pest and disease outbreaks, and facilitates rice-based multiple cropping.

④ Because the vegetative growth period is longer, it is advantageous for securing productive tillers. In addition, a higher optimal leaf area index enhances photosynthetic activity, and abundant solar radiation during the ripening stage increases the grain filling percentage and ultimately improves yield.

2) Late-Season and Late-Planting Cultures

(1) Late-Season Culture

① Late-season culture is practiced when transplanting is considerably delayed compared with normal-season planting, usually because of preceding crops in the cropping system or for pest and disease avoidance.

② In the southern plains of Korea, scheduled late-season culture is commonly adopted after harvesting potatoes or vegetables, resulting in late transplanting.

③ However, when both sowing and transplanting are delayed (late-season culture), the shortened growth duration leads to a substantial reduction in yield.

④ Therefore, varieties with low sensitivity to both temperature and photoperiod should be selected, and the practical limit for transplanting is from late June to early July.

(2) Late-Planting Culture

① Late-planting culture refers to cases where transplanting is unavoidably delayed due to drought, late harvest of the preceding crop, or water shortages, even though sowing was carried out at the recommended time.

② Because the nursery period is extended, the risk of premature heading increases.

③ In addition, seedlings tend to age in the nursery, so in unavoidable areas, late-planting-tolerant varieties should be chosen, and seeds should be sown sparsely in dry-bed nurseries to reduce competition among seedlings.

3) Double-Cropping

① Double-cropping refers to cultivating rice twice a year (two crop cycles) in the same paddy field.

② While double-cropping (two-crop) rice culture is common in subtropical regions, it is difficult to practice economically in Korea with current varieties and cultivation methods due to climatic limitations.

③ If double-cropping is attempted, it should combine extremely early planting with extremely late planting. The first crop uses an extremely early-maturing variety, which should be sown and raised in nurseries at an extremely early date, transplanted in mid- to late April, and harvested as early as possible. The second crop should use a photoperiod-sensitive and late-planting-tolerant variety, transplanted immediately after the first harvest and then grown to maturity.

④ Double-cropping can also refer to cultivating rice together with other crops in the same paddy field twice within a single year.

4) Reclaimed Saline Soil Culture

(1) Distribution and Characteristics of Reclaimed Soils

A: Distribution of Reclaimed Soils

① Korea has reclaimed large coastal areas, resulting in extensive reclaimed farmland in Jeonbuk, Jeonnam, Gyeonggi, Chungnam, and Gyeongnam Provinces.

② Because Korea has a small land area and is surrounded by the sea on three sides, future expansion of arable land must rely on reclamation projects.

B: Characteristics of Reclaimed Soils

① These soils generally have shallow effective soil depth and high salinity levels.

② Due to the influence of sodium (Na) and other salts, the soil particles are finely dispersed, resulting in poor physical properties.

③ Because the land is low-lying, the groundwater table is high, and saline water may seep to the surface.

(2) Salt Concentration and Salt Injury

① Rice is relatively tolerant to salt stress, and at concentrations below 0.1% there is little damage, while at 0.1~0.2% growth is somewhat inhibited.

② When salinity exceeds 0.3%, significant damage occurs. Thus, the threshold concentration for rice cultivation is below 0.3%.

③ Salt injury occurs most severely during the seedling establishment period immediately after transplanting and during tillering, rather than at the reproductive stage.

④ One of the mechanisms of salt injury is the accumulation of salts, particularly the direct effect of chloride (Cl^-), which reduces the functionality of chlorophyll.

⑤ Salt stress reduces enzymatic activity, inhibits photosynthesis, causes excessive nitrogen accumulation, delays growth and heading, and consequently lowers yield.

(3) Desalinization and Soil Improvement in Reclaimed Areas

A: Desalinization

① To cultivate rice on newly reclaimed land, the salinity must be reduced to below 0.3%.

② Frequent plowing and water replacement accelerate desalinization.

③ When plowing and irrigation are frequent, deep plowing is effective, and when irrigation is limited, shallow plowing helps reduce salinity in the root zone.

④ After transplanting, shortening the interval of water replacement enhances desalinization, and it is generally recommended to change the water every two days. However, frequent flushing results in substantial fertilizer loss, so the application rate should be increased accordingly.

⑤ epending on soil texture, it takes about 6 years in clay soils, 7 years in loam soils, and at least 2 years in sandy soils to reach the salinity level suitable for rice cultivation.

B: Subsurface Drainage and Desalinization

① To accelerate desalinization and prevent upward movement of salts from deeper layers, subsurface drainage should be installed from the early stages of reclamation.

② Subsurface drainage is essential for cultivating upland crops in reclaimed soils.

③ Without drainage, salinity reaches 0.3% in about 3 years, while with drainage at 8m spacing it takes 2 years, and with 5m spacing it takes more than 1 year.

C: Soil Improvement in Reclaimed Fields

① Reclaimed soils generally have high sodium and low calcium, high pH, and low organic matter. Incorporating rice straw, lime, and compost or applying topsoil improves soil fertility, enhances desalinization, and increases yields.

② Applying 300~400kg of lime during spring plowing is highly effective for desalinization and yield increase.

③ Reclaimed soils are often deficient in available zinc (Zn), which hampers rice growth. Applying 2.5kg of zinc sulfate per 10 ares improves plant growth.

(4) Rice Cultivation in Reclaimed Soils

① Salt-tolerant varieties such as Gyehwa-byeo, Gancheok-byeo, Seohae-byeo, Seomjin-byeo, and Yeongsan-byeo should be cultivated.

② In machine transplanting, seedlings should not be transplanted too late.

③ Because transplanting often results in gaps and missing hills, simultaneous rotary tillage and transplanting are recommended.

④ The soil structure is granular, and with high salinity the soil quickly hardens after puddling.

⑤ Since tillering is generally poor, planting density should be about 100 hills per 10 ares, with 5~6 seedlings per hill.

⑥ Fertilizer application should emphasize topdressing rather than basal application.

⑦ Because the soil is alkaline, ammonium sulfate (a physiologically acidic fertilizer) is recommended for nitrogen, superphosphate for phosphorus, and potassium sulfate (which suppresses sodium uptake) for potassium.

(5) Direct Seeding in Reclaimed Soils

① For direct seeding in reclaimed soils, broadcasting seeds on the flooded surface is recommended.

② In fields with low salinity (≤0.1%), plowing and rotary tillage before seeding is beneficial. In soils with moderate to high salinity (~0.3%), only rotary tillage should be done to minimize salt leaching.

③ The seeding rate varies with salinity but is generally 6~8 kg per 10 ares.

④ For water management, it is advantageous to drain water about 10 days after seeding, conduct soil compaction, and then re-irrigate 5 days later.

⑤ In saline water, the interval of water replacement should be every 2 days at high salinity and every 4 days at lower salinity.

⑥ Nitrogen fertilizer should be applied in split doses, while other fertilizers should be applied as basal dressing.

⑦ Weed flora vary with salinity: At very high salinity, species such as Suaeda, Salicornia, and Atriplex dominate, while at moderate salinity during early reclamation, weeds such as Scirpus and Aster tripolium are common, and at low salinity, typical paddy weeds such as Echinochloa, Monochoria, and Cyperus dominate.

5) Upland Rice Culture

(1) Upland Rice

① Rice cultivated in upland fields is also called yukdo (or sando). The distinction between lowland rice and upland rice is based on water conditions: rice grown under flooded conditions is lowland rice, while that grown without standing water is upland rice.

② In general, lowland rice can be grown in upland fields but yields are greatly reduced. Upland rice can also be grown in paddies with only small yield differences compared to upland conditions.

(2) Characteristics of Upland Rice

① Morphologically, upland rice has rougher leaves and stems and drooping, elongated leaves compared to lowland rice.

② It has a deeper root system with many fine roots, making it more tolerant to water deficit.

③ Physiologically, upland rice has higher oxygen demand in the soil than lowland rice and produces less sticky grains.

④ Because of lower water and silica uptake, upland rice is more susceptible to blast disease.

(3) Sowing and Cultivation Methods

① Upland rice is generally cultivated by dry direct seeding, and fertilizer application rates are about 30% higher than for lowland rice.

② Potassium (K) fertilizer should be applied at higher rates to enhance drought tolerance and lodging resistance.

③ Continuous cropping of upland rice depletes specific soil nutrients and reduces yields, so crop rotation is recommended.

6) No-Tillage Culture

① No-tillage culture refers to cultivating rice without plowing, usually by direct seeding.

② Advantages include saving labor and reducing costs associated with plowing machinery, thereby lowering production costs.

③ Compared with frequent plowing, it better maintains soil physical properties.

④ Disadvantages include difficulty in weed control, increased water requirement, lower fertilizer efficiency, and slightly reduced yield. While no-tillage reduces root cutting and may promote rice growth, it cannot provide weed control or soil physical improvement through tillage.

⑤ The use of herbicides for weed control adds extra cost.

⑥ No-tillage direct seeding reduces labor input and can lower methane emissions by up to 70%.

2. Direct Seeding Culture

1) Necessity and Types of Direct Seeding

① Direct seeding eliminates the need for nurseries, seedling boxes, and transplanting machines, thereby reducing input costs.

② In rice production, enhancing labor productivity and land productivity while reducing production costs to improve international competitiveness requires labor-saving cultivation technologies with minimal labor and input use, and direct seeding serves as a key method to achieve this.

③ Based on seeding methods, dry direct seeding, broadcast seeding, ridge-furrow direct seeding, and partial tillage direct seeding are distinguished.

④ Under flooded conditions, direct seeding methods include surface broadcasting, soil incorporation, ridge-furrow surface broadcasting, and no-till ridge direct seeding.

⑤ Dry direct seeding significantly reduces methane emissions compared with transplanting and is therefore considered an effective carbon reduction practice.

2) Characteristics of Direct Seeding

① Direct seeding saves labor and input costs but increases lodging susceptibility and weed infestation.

② Rice from direct seeding tends to produce lower-position tillers, which is advantageous for panicle number but may lead to overgrowth. In contrast, transplanted rice temporarily stops growing due to root cutting, preventing excessive tillering.

③ Germination and seedling establishment are often uneven, resulting in a low proportion of productive tillers.

④ Culm length and panicle length are shorter, stems are thinner, and roots are shallow, making plants more prone to lodging.

⑤ With the same sowing date, direct seeding usually advances heading compared to transplanting.

⑥ Although the number of panicles per unit area is higher, the number of spikelets per panicle is lower, resulting in somewhat lower yields than mechanical transplanting.

3) Yield and Labor-Saving Effects of Direct Seeding

① The yield of dry or wet direct seeding is slightly lower than that of conventional mechanical transplanting.

Table 8-2. Comparison of advantages and disadvantages between dry direct seeding and wet (Puddled) direct seeding

No.	Item	Dry direct seeding	Wet direct seeding
①	Aerial seeding	Inconvenient	Possible
②	Days to emergence	Relatively long (10~15 days on average)	Short (5~7 days on average)
③	Effort for soil pulverization	Much required	Less required
④	Field drainage	Requires careful management	No special concern
⑤	Germination rate	Poor	Good
⑥	Irrigation water requirement	Less	More
⑦	Land leveling (uniformity of field surface)	Difficult	Easy
⑧	Lodging degree	Less	More
⑨	Machinery operation	Easy	Difficult
⑩	Missing hills (poor stand establishment)	Rare	Frequent
⑪	Non-productive tillers (ineffective tillers)	Fewer	More
⑫	Position of tillering node	Higher	Lower
⑬	Rooting and establishment	Favorable	Poor
⑭	Sowing during rainfall	Difficult	Easy
⑮	Weed incidence	High	Low

② In dry direct seeding, yields may initially be lower but can become comparable as cultivation skills improve.

③ Production costs are somewhat reduced by both wet and dry direct seeding.

④ Labor input is reduced to 25% in wet direct seeding and 30% in dry direct seeding compared with machine transplanting.

(1) Problems and Countermeasures of Direct Seeding Cultivation

① Despite labor savings by eliminating seedling raising and transplanting, direct seeding has not been widely adopted because of unstable early seedling establishment, occurrence of weedy rice (red rice), and difficulties in weed control. As a result, its cultivation area remains below 1%.

② In 1995, the cultivated area under direct seeding reached 120,000 ha, exceeding 11%, but by 2021 it has almost disappeared.

③ In Japan, the area under direct seeding once declined, but puddled direct seeding has slightly increased again due to preference for easier practices associated with high-quality rice production.

④ Red rice (weedy rice) occurs frequently in dry direct seeding. Its seed coat is purple, it shatters easily, has strong germination capacity under low temperatures, and shows overwintering ability with wild rice characteristics.

A: Problems

① Low germination rate makes it difficult to secure sufficient seedling stand, leading to a reduction in the number of panicles.

② Shallow sowing depth increases lodging incidence.

③ Compared to transplanting, weed incidence is generally higher: about 2 times in puddled direct seeding and 3 times in dry direct seeding. In particular, weedy rice occurrence drastically increases in dry direct seeding.

B: Countermeasures

① To improve seedling establishment in puddled direct seeding, varieties with strong low-temperature germination ability and tolerance to low dissolved oxygen should be chosen. Careful land leveling and precise puddling operations are required.

② In dry direct seeding, if soil moisture is insufficient, germination is poor, and if it is excessive, field operations become difficult. Therefore, sandy loam or other suitable soils should be selected.

③ The optimum sowing depth is around 3cm. Herbicides are applied for weed control, but available herbicides are limited and their effectiveness is inconsistent.

C: Lodging reduction measures

① Lodging increases with higher seeding and fertilizer rates, as well as with continuous flooding. Therefore, thin seeding, reduced fertilizer application, split fertilizer application, and intermittent irrigation are recommended.

② Beginning 30 days after sowing, 2~3 mid-season drainings should be conducted. If excessive elongation raises the risk of lodging, growth retardants (e.g., lodging control chemicals) should be applied about 30 days before heading.

③ In puddled direct seeding, seeds should be sown immediately after puddling so that they are covered deeply enough to enhance stability.

(2) Direct Seeding and the Environment

A: Climatic conditions

① Since seeds are sown directly into the field without seedling raising, air temperature at sowing has a major effect on germination and seedling establishment.

② Early sowing prolongs the emergence period, resulting in poor stand establishment, while late sowing under high temperatures causes uneven emergence and delays heading, leading to yield reduction.

③ Sowing is recommended when daily mean temperature is above 12°C.

④ Rainfall has little effect on puddled direct seeding, but in dry direct seeding it is a decisive climatic factor for success or failure.

⑤ Excessive rainfall makes plowing, land preparation, and soil pulverization difficult, making sowing almost impossible.

⑥ Heavy rainfall after sowing submerges seeds, resulting in oxygen deficiency and poor germination.

⑦ Rain around the sowing period causes soil surface crusting, markedly reducing seedling establishment.

B: Soil conditions

① The success of direct seeding depends on soil properties and water management. Generally suitable sites include plains, flat lands, and well-irrigated paddies.

② Even in plains and well-irrigated paddy fields, direct seeding should be avoided in fields where cold underground water emerges, shaded valley bottoms, or saline soils with salt concentrations above 0.3%

③ In Korea, about 60% of the total paddy area is potentially suitable for direct seeding.

④ Dry direct seeding is suitable for sandy loam soils with good drainage.

⑤ If clay content is high, tillage and rotary operations (soil pulverization and leveling) become difficult, and rainfall severely hampers field work.

⑥ For clay-rich paddies, late fall or early spring plowing is helpful. During sowing, shallow ridges should be made for seeding, and drainage ditches should be established to maintain dry soil conditions.

⑦ For puddled direct seeding, silty loam or clay loam soils with moderate water-holding capacity are suitable.

⑧ In surface broadcasting under puddled conditions, excessive clay content causes seeds to be buried too deeply after puddling, reducing seedling establishment.

⑨ In reclaimed saline soils, puddled direct seeding often results in poor germination due to salt injury and oxygen deficiency caused by soil-reducing bacteria.

3. Environmentally Friendly Cultivation

1) Environmentally Friendly Agriculture

(1) Concept of Environmentally Friendly Agriculture

① Environmentally friendly agriculture, also referred to as environmentally conserving agriculture or low-input sustainable agriculture, is a farming system that simultaneously pursues productivity, resource and environmental conservation, and food safety.

② To achieve this, nutrient depletion caused by crops is compensated with organic matter or other natural inputs, while minimizing the use of artificial agrochemicals.

③ It relies on ecological nutrient cycling systems such as crop rotation and biological control, thereby conserving the agricultural environment and maintaining ecological balance.

④ By minimizing the application of pesticides and chemical fertilizers, this system secures stable productivity and profitability while preserving local resources and the environment, contributing to safer food production.

(2) Types of Environmentally Friendly Agriculture and Supporting Legislation

A: Organic agriculture

Organic agriculture excludes the use of chemical fertilizers, synthetic agrochemicals (pesticides, growth regulators, herbicides, etc.), and synthetic feed additives. Instead, only natural materials such as organic matter are used to produce crops and livestock products. It is particularly suited for small-scale farming.

B: Low-input agriculture

Low-input agriculture allows the use of chemical fertilizers, synthetic pesticides, and feed additives, but minimizes their application to reduce agricultural environmental pollution, conserve natural ecosystems, and produce safe agricultural products within permissible residue limits. In modern agriculture, low-input agriculture is often preferred over strictly organic farming.

C: Legislation for the promotion and management of environmentally friendly agriculture

① In Korea, environmentally friendly agriculture has been institutionalized based on the Environmentally Friendly Agriculture Promotion Act enacted in 1998.

② The purpose of this law is to enhance the environmental conservation function of agriculture and fisheries, reduce pollution caused by these sectors, foster farmers and fishers practicing environmentally friendly methods, and promote sustainable eco-friendly production. Furthermore, it aims to regulate eco-friendly agricultural and fishery products and organic foods, thereby protecting both producers and consumers.

③ Environmentally friendly agriculture and fisheries are supported as practices that either eliminate or minimize the use of synthetic pesticides, chemical fertilizers, antibiotics, and antimicrobial agents, thereby producing agricultural, fishery, and forestry products in a healthy environment.

D: Certification system for environmentally friendly agricultural and livestock products

① Environmentally friendly agricultural and livestock products refer to produce cultivated with no or minimal use of synthetic agrochemicals such as pesticides, chemical fertilizers, and feed additives, with the goal of conserving the environment and providing safer products to consumers.

② The certification system is a government-backed program where professional certification agencies rigorously evaluate and inspect these products based on strict criteria, and officially certify their safety and eco-friendly status.

(1) Organic Rice

A: Standards for organic rice production

① Cultivation must be conducted for more than three years in fields managed through crop rotation or the application of organic fertilizers, without the use of chemical fertilizers or synthetic pesticides. During production, processing, and distribution, no toxic or hazardous substances should be used.

② Residual pesticide levels must not exceed one-twentieth of the legal tolerance limit.

③ For quality certification, paddy soils must not exceed the contamination thresholds set by the Soil Environment Conservation Act, and the fields must have followed the prescribed organic cultivation practices for at least three years prior to the first harvest.

④ Irrigation water must meet or exceed agricultural water quality standards as defined in the Framework Act on Environmental Policy and the Groundwater Quality Conservation Act.

⑤ Seeds must, in principle, be "organic seeds" produced and managed in accordance with organic certification standards, and genetically modified (GMO) seeds are prohibited.

Table 8-3. Types and standards of environmentally friendly agricultural and livestock products

No.	Certification type	Certification standards	Main requirements
①	Organic agricultural products	There is no use of pesticides and fertilizers, and crop rotation (sequential cropping) must be prioritized for nutrient and pest/disease management. Agricultural by-products should be returned to the soil, and other substances permitted in organic farming may be used for nutrient supply and pest control, according to prescribed purposes and conditions.	Emphasis on rotation, soil fertility management, and natural input use.
②	Organic livestock products	Animals must be fed organic feed, and barns and pastures must meet minimum space requirements by species to allow free movement. Disease prevention is achieved through proper breed selection and hygienic management of facilities. Veterinary drugs cannot be used in the absence of disease. Integrated with the former "antibiotic-free livestock" certification.	Organic feed, adequate space, preventive health management, no routine antibiotics.
③	Organic apiculture products	Within a 3km radius of hives, nectar sources must be uncontaminated and meet organic standards. Synthetic pesticides, veterinary drugs, and other prohibited substances cannot be used for bee disease/pest control. Beehives must be constructed from natural materials.	Clean nectar environment, no synthetic chemicals, natural hive materials.
④	Organic processed foods	Must be manufactured/processed using certified organic agricultural products. Records of raw material purchase/use and production/sales of organic processed foods must be maintained. Only additives permitted for organic processing may be used, in accordance with intended purposes and conditions.	Traceability, certified inputs, limited additives.
⑤	Non-food organic processed products (organic feed for livestock and companion animals)	Manufactured/processed using organic agricultural and livestock products, organic processed foods, and approved single or supplementary feeds. Records of raw material purchase/use and production/sales must be maintained. Only additives permitted in organic feed production may be used.	Organic-based feed, approved supplements, strict record-keeping.
⑥	Pesticide-free agricultural products	No pesticides are used, and fertilizers are limited to one-third or less of the recommended application rate. Agricultural by-products must be returned to the soil, and only substances permitted in organic farming may be used for nutrient supply and pest/disease control. Certification for "low-pesticide products" is not implemented.	No pesticides, minimal fertilizer, permitted natural inputs.
⑦	Processed foods using pesticide-free raw materials	Must be manufactured/processed using certified pesticide-free agricultural products. Records of raw material purchase/use and production/sales must be maintained. Only additives permitted in organic food processing may be used, in accordance with purposes and conditions.	Certified pesticide-free inputs, traceability, limited additives.

*Source: National agricultural products quality management service (NAQS), www.enviagro.go.kr

B: Organic rice cultivation methods

① No chemical fertilizers or synthetic pesticides are permitted. Long-term rotation with legumes, green manure crops, or deep-rooted crops must be practiced.

② Organic matter applied to soils must be produced according to organic certification standards.

③ Livestock manure must be fully composted and applied without causing environmental pollution from excessive use, runoff, or leaching.

④ Pest and weed control must be carried out through varietal selection, appropriate crop rotation, mechanical tillage, utilization of natural enemies, mulching, mowing, flame weeding, or animal grazing.

C: Quality control of organic rice products

① Radiation cannot be used for pest control or food preservation. Packaging materials must comply with the Food Sanitation Act and, preferably, should be biodegradable, recycled, or recyclable.

② Residual pesticides are permitted only in unavoidable cases, such as contamination by drift from neighboring conventional farms, irrigation water, or drainage from adjacent fields, and the allowable level must not exceed one-twentieth of the maximum residue limit (MRL) for agricultural products as stipulated by the Ministry of Food and Drug Safety (MFDS).

(2) Pesticide-Free Rice

A: Standards for pesticide-free rice production

① Rice produced without any use of synthetic pesticides and with chemical fertilizers applied at no more than one-third of the recommended rate, with residual pesticide levels not exceeding one-twentieth of the legal tolerance limit.

② Herbicide use is strictly prohibited.

③ For quality certification, paddy soils must not exceed contamination thresholds set under the Soil Environment Conservation Act.

④ Irrigation water must meet or exceed agricultural water standards under the Framework Act on Environmental Policy and the Groundwater Quality Conservation Act.

⑤ Organic seeds must be used, and genetically modified (GMO) seeds are prohibited.

B: Cultivation methods for pesticide-free rice

① Chemical fertilizers must be applied at no more than one-third of the nutrient levels recommended for each cultivation field by the administrator of the rural development administration, the director of the agricultural research and extension services, or the head of the agricultural technology center, and no synthetic pesticides shall be used.

② Long-term rotation with legumes, green manure crops, or deep-rooted crops must be implemented.

③ Livestock manure must be fully composted, and excessive use, runoff, or leaching that could cause environmental pollution must be avoided.

④ Pest and weed control should rely on varietal and crop selection, proper rotation, mechanical tillage, natural enemies, mulching, mowing, flame weeding, and animal grazing.

C: Quality control of pesticide-free rice products

① Preventive measures must include removal of pest habitats, facility barriers, and the use of mechanical, physical, or biological methods. If these are insufficient, only environmentally approved inputs may be used for pest control. Radiation is not allowed.

② Residual pesticides shall be permitted only in unavoidable cases, such as contamination caused by drift from neighboring conventional farms, irrigation water, or drainage from adjacent fields, and the allowable level must not exceed one-twentieth of the maximum residue limit (MRL) for agricultural products as stipulated by the Ministry of food and drug safety (MFDS).

Table 8-4. Types of environmentally friendly rice

No.	Item	Organic rice	Pesticide-free rice
①	Residual pesticide levels	≤ 1/20 of the maximum residue limit (MRL)	Prohibited
②	Use of chemical fertilizers	Prohibited	≤ 1/3 of the recommended application rate
③	Use of genetically modified (GMO) seeds	Prohibited	Prohibited
④	Use of herbicides	Prohibited	Prohibited
⑤	Use of radiation	Prohibited	Prohibited
⑥	Use of synthetic pesticides	Prohibited	Prohibited

③ If pesticide residues are detected for which no MRL has been established, the detected amount must not exceed ≤ 1/20 of the provisional MRL set by MFDS.

(3) Comparison between Organic Rice and Pesticide-Free Rice

① Since the abolition of the "Low-Pesticide Certification System" in 2016, environmentally friendly rice has been classified into organic rice and pesticide-free rice. Both prohibit the use of synthetic pesticides, herbicides, radiation, and genetically modified (GMO) seeds.

② Residual pesticide levels must not exceed one-twentieth of the legal tolerance limit. The key distinction is that organic rice prohibits the use of chemical fertilizers entirely, whereas pesticide-free rice allows limited application of chemical fertilizers up to one-third of the recommended rate.

3) Considerations in the Production of Environmentally Friendly Rice

(1) Organic and Chemical Fertilizers

① It is not necessarily correct to assume that organic fertilizers are always good and chemical fertilizers are always bad. Likewise, environmentally friendly farming does not always guarantee beneficial effects on the environment and ecosystem.

② Over-application of organic fertilizers can have the same negative effects as excessive chemical fertilizer use, such as soil salinization and water pollution.

③ Organic fertilizers must be decomposed by soil microorganisms into inorganic forms before plants can utilize them.

④ Excessive application of organic fertilizers increases nitrogen uptake, raising grain protein content and reducing eating quality (palatability).

⑤ Rational application of chemical fertilizers, based on soil testing and standardized recommendations, imposes little environmental burden and supports healthy plant growth without reducing eating quality.

(2) Environmentally Friendly Weed Control

① Weed control is the most challenging issue in environmentally friendly rice production because certification for both organic and pesticide-free rice strictly prohibits herbicide use. Success therefore depends on weed management.

② Alternatives include introducing ducks, golden apple snails (*Pomacea canaliculata*), rice bran, freshwater crabs, or loaches to paddy fields for weed suppression.

(3) Pest and Disease Control and Pesticide Use

① Cultivating rice without any pesticide assistance is extremely difficult. Farmers must start by selecting resistant varieties and maintaining robust plant growth while avoiding practices that reduce genetic resistance, such as excessive fertilization.

② Modern pesticides generally have low toxicity and short residual periods, and thus if usage guidelines are followed, risks are minimal. Residue inspections are conducted both during production and at postharvest stages, including storage and distribution.

③ Maximum residue limits (MRLs) are set at levels considered safe for lifetime daily intake, based on recommendations from the FAO/WHO Joint Meeting on Pesticide Residues (JMPR).

④ In Korea, the "Guidelines for the Safe Use of Pesticides" are enforced under the Pesticide Control Act. These specify target pests, timing of application, and preharvest intervals, as set by the Rural Development Administration, and must comply with the pesticide residue standards announced by the Ministry of Food and Drug Safety (MFDS).

⑤ The National Agricultural Products Quality Management Service (NAQS) conducts safety inspections during production, storage, and shipment under the Agricultural and Fishery Products Quality Control Act. Safety of distributed and imported agricultural products is examined by the MFDS under the Food Sanitation Act.

⑥ The Positive List System (PLS) for pesticide residues has been implemented since 2019, and compliance is mandatory.

(4) Environmentally Friendly Agricultural Inputs

① Production of environmentally friendly rice requires organic fertilizers and eco-friendly agricultural materials such as bio-pesticides and bio-fungicides.

② Standards for the use of such materials are regulated by the Act on the Promotion of Environmentally Friendly Agriculture and Fisheries and the Management and Support for Organic Food, etc. under the Ministry of Agriculture, Food and Rural Affairs (MAFRA).

③ When using livestock manure as an organic fertilizer, manure from factory-type farms that rely on non-compliant feed or veterinary drugs cannot be used.

4) Principles of Environmentally Friendly Rice Production

Safe rice production begins with the selection of varieties that are genetically resistant to pests and diseases. Maintaining environmentally friendly cultivation is difficult with high-yielding varieties that lack such resistance.

(1) Low-Input Farming Practices

① When applied at appropriate levels, organic fertilizers, chemical fertilizers, and pesticides do not cause serious problems for the environment or food safety.

② Environmentally friendly agriculture should begin with the low input of agricultural materials to promote healthy plant growth. (a) Integrated Pest Management (IPM) reduces pesticide use. (b) Integrated Nutrient Management (INM) applies optimization techniques to reduce chemical fertilizer use. (c) Recently, precision agriculture technologies have been developed, in which tractors and drones linked with Geographic Information Systems (GIS) and satellites apply site-specific optimal amounts of fertilizer to each paddy field.

(2) Weed Management

A: Reducing Weed Seed Density

① Weeds produce a massive number of seeds that remain buried in the soil. These seeds can stay dormant for years and germinate once conditions are favorable.

② Effective weed management without herbicides requires preemptive reduction of soil weed seed banks.

B: Creating an Environment Unfavorable to Weeds

① Many weed seeds are light-sensitive, particularly responsive to red wavelengths.

② Shading the soil surface reduces weed emergence.

③ Duck farming systems not only allow ducks to feed on weeds and seedlings but also create turbidity in water, reducing light penetration.

④ Application of rice bran or fermented oilseed cake is also effective, and when combined with rice straw or grasses, these promote algal growth that blocks light and suppresses germination.

⑤ Rapid decomposition of organic matter by aquatic microbes can temporarily deplete oxygen and produce gases such as methane, which weaken weeds.

⑥ Physical control methods include hand-weeding, traditional weeders, or flame torching. In all cases, preventive management to reduce weed density is critical.

C: Integrated Weed Management

① To enhance weed control effectiveness, multiple strategies must be combined.

② Removing mature weed seeds immediately after heading prevents seed dispersal for the next season.

③ Soil tillage and water management help reduce weed seed density in advance.

④ After transplanting, field conditions should be managed to disadvantage weeds.

⑤ Emerged weeds can be removed by flame torching, mechanical cutting, or manual weeding.

⑥ In organic and eco-friendly farming, the concept is weed suppression rather than weed eradication, as complete elimination is impractical and uneconomical.

⑦ In Korea, where farm sizes are relatively small, environmentally friendly farming often requires labor-intensive management. For small-scale farms seeking high returns, production can become high-cost.

4. Cropping Systems

1) Transition of Cropping Systems in Paddy Fields

(1) Cropping System

① A cropping system refers to the sequence or arrangement of different crops in a given field, designed to increase land-use efficiency, maintain and improve soil fertility, enable sustainable agriculture, and raise farm income.

② An ideal cropping system divides farmland into 3~5 parts, alternates cereals and legumes, and occasionally introduces soil-improving crops or root crops that promote subsoil improvement.

③ In Korea, due to limited land resources, such systems are difficult to practice, and under flooded paddy conditions the cropping system is particularly restricted.

(2) History of Cropping Systems in Korea

① Since the 15th~17th centuries, Korea has primarily practiced a simple rice–barley double-cropping system.

② Until the mid-20th century, about 10~20% of paddies (mainly in southern regions) adopted double cropping.

③ During the rapid economic growth era (1960s~1970s), double cropping was practiced on about 80% of paddy land.

④ In the 1970s, the rice–barley system was replaced with economic crops such as vegetables and minor cereals.

⑤ In the 1980s, the rice–barley system further declined, while cultivation of vegetables and non-cereal economic crops increased significantly.

⑥ In the 1990s, vegetables, potatoes, and forage crops expanded further while rice–barley systems almost disappeared.

⑦ Since the 2000s, wheat has virtually disappeared, and only limited areas grow barley, forage crops, or horticultural crops.

2) Rational Cropping Systems in Paddy Fields

① A rational cropping system must maintain farm income while improving soil fertility, ensuring low-input yet sustainable agriculture.

② A common method is to grow rice in summer and cultivate winter crops such as cereals or green manure crops.

③ Through rice-upland rotation (dapjeon yunhwan), rice is followed by upland crops, which enhances fertilizer efficiency and pest/disease control.

④ As double-cropping (daprijak), winter crops such as hardy cereals, horticultural crops, or green manure crops are cultivated after rice harvest.

⑤ Although limited, in central and northern Korea rye is suitable as a forage crop due to its cold hardiness, while in southern regions Italian ryegrass with high forage value is recommended.

⑥ For soil fertility improvement, leguminous green manure crops such as Chinese milk vetch (Astragalus sinicus) and hairy vetch (Vicia villosa) are advantageous.

(1) Major Double-Cropping Crops

A: Rye (*Secale cereale*)

① Rye grows well even in poor soils and harsh environments, and its strong cold tolerance makes it suitable as a winter forage crop in central and northern Korea. It is sown in autumn after rice harvest and cut in spring for silage or green fodder.

② Early sowing increases yield. In the case of very early sowing, grazing or green cutting is possible before overwintering. Sowing is done in October, ensuring at least 4 true leaves before winter.

③ A simplified method involves rotary tillage, followed by broadcasting seeds and fertilizer, then light rotary tillage again.

④ The recommended fertilizer application rate is 20 kg of nitrogen (N), 12 kg of phosphorus (P_2O_5), and 12 kg of potassium (K_2O) per 10 ares, with careful attention to drainage.

⑤ Fresh matter yield peaks at the heading stage and declines after flowering, so for green forage, harvesting between heading and flowering is ideal.

⑥ Dry matter yield is maximized at the milk stage (early grain formation), so harvest between flowering and milk stage for silage production.

B: Chinese Milk Vetch (*Astragalus sinicus*)

① As a green manure crop, sowing is recommended in early September at 3~4 kg per 10 ares.

② When sown under flooded conditions during rice seedling establishment, late sowing leads to poor growth and reduced overwintering ability. Fields should be drained within 10 days after sowing, since too early drainage reduces rice yield and quality, while too late drainage suppresses milk vetch growth.

③ If growth is poor, a small amount of nitrogen fertilizer can be applied in spring.

Phosphorus and potassium fertilizers are highly effective and should be applied soon after rice harvest.

④ Harvest milk vetch about 10 days before rice transplanting. Stop irrigation, incorporate into dry soil with rotary tillage, and re-flood after 5 days to reduce organic acid release and enhance rice establishment.

C: Hairy Vetch (*Vicia villosa*)

① When cultivated as a green manure crop, the appropriate sowing time is October with a seeding rate of 6~9 kg per 10 area, and earlier sowing increases green manure biomass production.

② Sowing can be done by broadcasting before rice harvest or immediately after harvest with rotary tillage. Under standing rice, broadcasting is done from about 10 days before harvest. This reduces labor and advances sowing time.

③ Hairy vetch is incorporated into the soil 2~3 weeks before rice transplanting at a rate of 1,500~2,000 kg per 10 ares.

④ When used as green manure, no nitrogen fertilizer is needed, while phosphorus and potassium fertilizers should be applied at standard rates.

(2) Effects of Paddy-Upland Rotation Culture

① Soil cracks develop deeply in both clay loam and sandy loam soils, and surface soil becomes aggregated, breaking the hardpan.

② Soil aeration and porosity are improved.

③ Soil chemistry improves, with increases in exchangeable calcium and magnesium cations.

④ Weed populations, nematodes, and soil-borne pests/diseases are reduced, enhancing crop productivity.

Rice Crop Science

Chapter 9. **Quality and Post-Harvest Management**

Chapter 9.
Quality and Postharvest Management

Even if agricultural products are produced under optimal conditions with advanced farming techniques and improved varieties, they become meaningless if they fail to gain consumer acceptance. This chapter focuses on milling, processing, and utilization of rice as the final product, and aims to broaden understanding as a foundation for further advancement. Although many aspects are still in progress, diverse technologies have been integrated, improving convenience in distribution and utilization. Rice is more than food, as it is a unique product embedded with cultural heritage. Here, the overall development of rice processing technologies is reviewed, along with ideas for future improvement.

1. Postharvest Management

Postharvest management refers to the processes of drying, storage, milling, and distribution of rice, and the related practices are called postharvest technologies. In the era of high-quality agricultural production and liberalization of rice trade, postharvest management has become increasingly important.

1) Drying of Rice

(1) Principles of Drying

① Rice must be dried promptly after harvest to ensure safe storage. High moisture content promotes microbial growth, internal biochemical changes, and enzymatic activity, all of which accelerate quality deterioration.

② Moisture content of freshly harvested rice is generally around 25%, but may be as low as 15% when harvested late or as high as 40% at maximum.

③ For favorable storage, rice should be dried to about 15% moisture. Below 15%, storability increases but eating quality deteriorates.

④ Rapid drying of rice induces a high incidence of fissured kernels and other defects, leading to quality deterioration, and therefore gradual drying is required.

(2) Drying Methods

A: Mechanical drying

① Mechanical drying involves harvesting rice with a combine, threshing simultaneously, and drying with a dryer. Methods include heated-air drying (hot-air drying) and ambient-air drying (improved bin drying). This is the most widely used method today.

Figure 9-1. Photograph of mechanical and sun drying of rice

B: Sun drying

① In the past, rice cut with a sickle was spread on the paddy surface to dry, or bundled into small sheaves and stood upright in the field, but this practice is no longer seen today.

② Once rice is partially dried, it is threshed with a thresher and then subjected to final drying for storage.

③ Final drying methods include drying on concrete floors, asphalt surfaces, or nets. The optimal grain layer thickness for drying is about 5cm.

④ When rice is thinly spread on asphalt surfaces, rapid drying increases fissured kernels, lowering grain quality.

⑤ In improved bin drying, harvested rice is placed into the bin and aerated until moisture content reaches 15%, after which it can be stored directly while maintaining quality and reducing costs.

(3) Drying Technologies

A: Moisture content

① Rice should be dried below 45°C and maintained at around 15% moisture.

② When the moisture content falls below 14%, storability improves but eating quality declines, and when it exceeds 16%, milling efficiency and palatability increase but the grain becomes more susceptible to deterioration.

B: Drying temperature

① For hot-air dryers, an optimal drying temperature is around 45°C. At this temperature, milling recovery and germination rates are high, fissuring and breakage rates are low, and drying time is reasonable (~6 hours).

② When rice is dried at high temperatures above 55°C, the rates of fissuring and breakage increase sharply. The higher the drying temperature exceeds 55°C, the greater the fissuring, accompanied by protein denaturation and starch retrogradation, which reduce germination ability and lower eating quality by decreasing stickiness during cooking.

③ Rice with lower initial moisture suffers less eating quality reduction under high-temperature drying, but rice with high moisture loses quality significantly when dried at high temperatures.

C: Heating conditions and drying rate

① When using hot-air dryers, a heating rate of about 1°C per hour is appropriate. Rapid high-temperature drying causes a sharp increase in fissuring and denaturation of organic compounds, leading to quality loss.

② The drying rate should be about 1% moisture reduction per hour.

③ To suppress fissured kernels, the higher the initial grain moisture, the lower the starting air temperature should be. When humidity is low, the humidity of the drying air should also be kept low.

D: Drying and quality deterioration factors

① Rapid drying causes a large occurrence of fissured kernels, lowering quality. Fissured rice produces a rough texture and leaches starch during cooking, negatively affecting palatability.

② The reason fissuring increases during rapid drying is that moisture does not evaporate uniformly from the grain surface, causing cracking.

③ Improper drying methods (overheating, overdrying, rapid drying) result in fissured grains, off-flavors, and discolored kernels, all lowering quality.

④ Delayed drying accelerates deterioration, especially in high-moisture rice.

⑤ Excessive high-temperature drying not only reduces eating quality but also decreases milling efficiency due to low grain moisture and produces heat-damaged kernels.

2) Storage

① During storage, rice consumes its reserves through respiration, while microorganisms and insects proliferate. This process generates heat and moisture, accelerating deterioration and quality loss.

② Storage is as important as production. To minimize quantitative losses and compositional changes, proper storage conditions and facilities are required.

(1) Physicochemical and Biological Changes During Storage

① Long-term stored rice shows overall decline in color, stickiness, taste, and aroma, leading to reduced eating quality.

② Lipid oxidation causes rancidity, increasing free fatty acids and producing off-odors.

③ Acid value, measured as mg KOH required to neutralize free fatty acids per 100g dry matter, is an indicator of storage condition, and values above 20 indicate the onset of spoilage.

④ Starch is degraded by α-amylase, increasing reducing sugar content.

⑤ Respiration and moisture loss reduce grain weight, and storage losses average about 5%.

⑥ Germination ability declines, and after more than four years of storage, seed viability decreases sharply.

(2) Factors Affecting Storability

① Fungal development occurs readily when relative humidity exceeds 70% and temperature is above 22 °C. Fungi consume proteins and carbohydrates, resulting in both quantitative and qualitative losses. Their incidence increases rapidly when the relative humidity inside storage facilities reaches around 80%.

Table 9-1. Comparison of storage characteristics according to rice grain form

No.	Item	Rough rice (paddy)	Brown rice	Milled rice (white rice)
①	Degree of color change	Minimal	Moderate	High
②	Germinability	Viable	Viable	Non-viable
③	Grain form	The whole grain including the husk.	Rough rice with the husk removed.	Brown rice from which the aleurone laye has been removed.
④	Hygroscopicity	Low	Moderate	High (sensitive)
⑤	Pest/disease resistance	Strong	Moderate	Weak
⑥	Postharvest compositional change	Minimal	Moderate	High
⑦	Storability at ambient temperature	High	Moderate	Low
⑧	Storage characteristics	It has higher storability than brown rice and milled rice, and is safer against pests and diseases.	It occupies less volume than rough rice, thereby reducing warehouse space requirements and packaging/distribution costs.	It is highly sensitive to external temperature and humidity, making it prone to deterioration, more vulnerable to insect damage, and reduced in eating quality.
⑨	Storage method	Korea (requires about twice the warehouse space, but due to low humidity storage is possible without special control facilities).	Japan (requires less warehouse space, but because of high humidity rice must be stored in low-temperature controlled facilities).	Ordinary households (in long-term storage, pest infestation and odor may occur, and eating quality as well as overall grain quality are easily deteriorated)
⑩	Storage volume	Largest	~55% of rough rice volume	~50% of rough rice volume
⑪	weight	Heaviest	~80% of rough rice weight	~75% of rough rice weight

② Moisture conditions: Safe storage can generally be maintained even under hot and humid summer conditions if the grain moisture content is kept at approximately 15%.

③ At storage temperatures above 15 °C, insect pests such as rice weevils (*Sitophilus oryzae*) and storage moths proliferate in the bran and embryo layers, but polished white rice, from which the bran and embryo are removed, is rarely infested.

④ Oxygen is essential for respiration, but lowering its concentration reduces respiratory consumption and spoilage. Injecting carbon dioxide (CO_2) or nitrogen (N_2) to lower oxygen (O_2) levels is effective in preventing pests and diseases during storage.

(3) Storage Facilities and Technologies

A: Rice storage facilities

① In Korea, storage warehouses are well equipped with temperature and humidity control systems. Rice is stored in the rough (paddy) form, allowing for high storage capacity.

② In the United States, most rice is stored in large silos with low-temperature storage systems, separated by variety, production region, and grade, to maintain storage quality.

③ In Japan, due to high humidity, rice is stored in the brown rice form.

B: Rice storage conditions

① At ambient temperature storage, the quality of milled rice and brown rice, as well as even rough rice, deteriorates significantly during summer because of respiratory consumption and the occurrence of storage pests.

② Storing rice below 15°C prevents the occurrence of storage pests during summer and suppresses quality deterioration such as an increase in free fatty acids, thereby improving storability.

③ For brown rice storage, when moisture content exceeds 20%, storage below 10°C is appropriate, and when moisture content is around 15%, storage at 15°C is preferable.

④ For long-term safe storage, moisture content should be maintained around 15% and relative humidity around 70%.

⑤ The storage atmosphere should be controlled to approximately 5% oxygen (O_2) and 3% carbon dioxide (CO_2).

⑥ Since rice has a relatively low moisture content of around 15%, unlike fresh vegetables or fruits with approximately 90% moisture content, it does not need to be stored at near-freezing temperatures (around 0°C).

2. Rice Milling

(1) Fundamentals of Milling

A: Principle of milling

① When brown rice is consumed without milling, it is coarse, has poor taste, and shows low digestibility.

② Milling is the process of removing the bran layer from rice grains, which are protected by the hull (pericarp) and the seed coat (testa). These protective tissues contain components and structures that are difficult to digest, and milling improves the edibility and processability of rice.

B: Terminology of milling

① Unhulled rice, also called paddy or rough rice, refers to the grain that still contains the hull (pericarp), the bran layer (testa), the embryo, and the endosperm.

② Brown rice, also called hulled rice, refers to the grain in which only the hull has been removed.

③ Milled rice, also called white rice, refers to the grain obtained when the bran layer, including the testa and aleurone layer, is removed from brown rice.

④ Milling refers to the process of producing milled rice by removing the hull and bran layers from paddy, during which by-products such as chaff (rice husk), rice bran, and broken rice are generated.

C: Milling ratio

① The milling ratio refers to the proportion of milled rice to rough rice, which is calculated as (dehulling rate × whitening rate) divided by 100, and is typically around 70 to 74 percent.

② The dehulling rate refers to the proportion of brown rice that does not pass through a 1.6 millimeter sieve after the hull is removed. Although this varies depending on the variety, about 80 percent of paddy is converted into brown rice.

Table 9-2. Degree of milling and removal of the bran layer

No.	Degree of milling	Extent of bran layer removal
①	5-bu milled rice (5-part polished rice)	• Milled so that 97% of the brown rice weight remains. • About 50% of the bran layer to be removed is eliminated. • The embryo (germ) remains.
②	7-bu milled rice	• Milled so that 95% of the brown rice weight remains. • Bran layer on the ventral side of the grain is completely removed.
③	9-bu milled rice	• Milled so that 92.9% of the brown rice weight remains (7% removed). • Bran layers on the dorsal and upper sides of the grain are completely removed. • 100% of the bran layer to be removed is eliminated.
④	12-bu milled rice	• Milled so that 90.4% of the brown rice weight remains (10% removed). • All bran layers of the grain are completely removed.

*Note: This has been the official standard for milled rice yield since 2020.

③ The whitening ratio refers to the percentage of brown rice converted into milled rice. For 9-bu milled rice, the whitening ratio is 92.9%, and for 12-bu milled rice it is 90.4%.

④ The milling loss ratio indicates the percentage of weight loss during milling, such as rice bran and embryo separated from brown rice. In 5-bu milled rice, the milling loss is about 4%, while in 10-bu milled rice it is about 8%.

D: Factors affecting milling loss

① The thickness of the bran layer affects the loss, and the proportion of bran to the whole grain is lower in large-grain varieties than in small-grain varieties and lower in fully ripened rice than in immature rice, resulting in smaller milling loss.

② Grain drying conditions influence the loss, as well-dried kernels produce lower milling loss, while grains exposed to rain during drying or subjected to repeated cycles of moisture absorption and drying show greater milling loss.

③ The milling method is also important, and removing small amounts gradually through repeated light milling reduces milling loss.

④ The timing of milling matters, since milling immediately after harvest combined with sufficient drying results in lower milling loss.

E: Determination of milling degree

① Methods of determining the degree of milling include color-based judgment, milling time, number of milling passes, electric power consumption, degree of bran removal, MG staining, and ME staining methods.

② Since visual judgment of milling degree based on grain color can produce large errors due to varietal differences and light effects, staining methods reduce error. The widely used ME staining method evaluates the degree of bran removal using a mixture of methylene blue and eosin Y, and the iodine staining method may also be used as a supplementary approach.

③ The whiteness meter method is more practical, and although staining methods are accurate, they are relatively complicated. Therefore, the degree of milling is expressed as a whiteness value measured by a whiteness meter, and at Rice Processing Complexes (RPCs), a whiteness value of 40 is typically used as the standard.

(2) Milling Process

A: Milling action

① Milling is a process in which friction, abrasion, cutting, and impact act in combination.

② The fundamental mechanisms of milling are friction and impact, accompanied by abrasion and cutting as intermediate actions, resulting in a comprehensive effect.

③ Rice milling machines are generally classified into friction-type and abrasive-type milling machines.

B: Milling stages

① The milling process begins with raw paddy, followed by cleaning, dehusking, separation of brown rice, whitening, and separation of broken rice, and it ends with the production of milled rice.

② One of the most important factors affecting milling is the moisture content of the raw paddy, and the optimum moisture content for milling is about 16%.

③ The lower the moisture content, the greater the electrical energy required.

④ During the milling of rice, approximately 0.3~0.5% of moisture is lost through evaporation, so humidification may be necessary.

Table 9-3. Terminology related to rice milling

No.	Category	Term	Definition
①	Grain (rice kernels)	Whole milled rice grain (head rice)	Milled rice remaining on a 1.7mm sieve. From 100g of these grains, only fully intact kernels are included, excluding damaged, discolored, mixed, immature, and chalky grains. Broken kernels are included only if they are ≥ 3/4 of the average length of a whole grain.
		Chalky grain	A kernel in which more than half of the volume is chalky.
		Normal grain	A kernel excluding damaged, immature, discolored, aborted, mixed, or foreign kernels.
		Damaged grain	A kernel contaminated or deteriorated, including sprouted, diseased, insect-damaged, decayed, spotted, or blackened grains. Minor damage not affecting quality is excluded.
		Discolored grain	A kernel with all or part of the surface stained brown or red. Minor discoloration not affecting quality is excluded.
②	Broken and mixed grains	Mixed grain	Grains of crops other than rice.
		Small broken rice	Fragments that pass through a 1.7mm sieve but not through a 1.4mm sieve.
		Large broken rice	Fragments that do not pass through a 1.7mm sieve and are shorter than half the average length of a whole grain.
③	Ratios	Immature grain ratio	The proportion of small, immature grains separated after husking 1kg of paddy and sieving through a 1.6mm sieve.
		Broken rice ratio	The percentage of fragments passing through a 1.4mm sieve, expressed as a percentage of the brown rice used.
		Dehulling ratio	The percentage of brown rice that does not pass through a 1.6mm sieve after husking, expressed relative to paddy weight.
		Whitening ratio	The percentage of milled rice obtained by milling 1kg of brown rice and sieving through a 1.4mm sieve, expressed relative to brown rice weight.
④	Foreign materials	Stone	Mineral solids such as stones or concrete fragments remaining on a 1.4mm sieve.
		Foreign matter	In paddy: anything other than rice kernels. In brown rice: non-rice materials or particles passing through a 1.7mm sieve. In milled rice: particles passing through a 1.4mm sieve.
⑤	Weight	Test weight (bulk density)	The weight of 1 liter of rice, usually measured with a Burrell-type grain tester.

Table 9-4. Facilities of rice processing complexes (RPCs) for high-quality rice production

No.	Equipment	Description
①	Bran removal machine for brown rice	A machine that removes bran from brown rice. For conditioning (tempering), the optimal humidification rate is about 0.2% of the weight per hour for milled rice and about 0.3% for brown rice at a grain temperature of 25°C.
②	Brown rice separator	A machine that separates brown rice from the mixture of brown rice and unhulled rice produced during the first husking stage (approximately 86% efficiency).
③	Color sorter	A device that removes foreign materials, discolored grains, and malformed grains.
④	Grading sorter	Equipment that removes broken rice from milled rice to produce head rice (whole kernels).
⑤	Green Active system	Equipment that improves rice quality and increases milling yield by stabilizing and activating the physical properties of rice.
⑥	Moisture controller	A device that adjusts the moisture content of low-moisture rice to an optimal level above 15%.
⑦	Rice milling machine (polisher)	When rice with different moisture contents is mixed, moisture transfer occurs between grains, and after one day of mixing, equilibrium is nearly reached, so final milling is performed two days after mixing.
⑧	Wet polisher	Equipment for producing "clean rice" that does not require washing before cooking.

Figure 9-2. Photograph of a color sorter in operation

Figure 9-3. Photograph of a brown rice separator in operation

Figure 9-4. Rice processing complex (RPC) overview

(3) Milling and Quality

A: Paddy rice

When rice is stored in the un-milled (paddy) form, deterioration progresses slowly. However, once brown rice is milled into white rice, deterioration accelerates. Milling damages the cells of the bran layer, causing lipid oxidation in the cell membranes, which results in rancidity and loss of eating quality.

B: Milled rice (White rice)

Rice is a living collection of cells, and over time its respiratory energy is consumed, leading to reduced palatability. Therefore, rice should be consumed within 1~2 weeks after milling. In winter, when temperatures are low, eating quality can be maintained for up to one month, whereas in summer, rice should be consumed within seven days after milling to avoid quality decline.

(4) Packaging of Rice

A: Packaging materials

① In the past, straw bags were used, but currently jute bags, paper bags, polyethylene (PE) film, polypropylene (PP) bags, and metal-coated packaging materials (aluminum, nickel, chromium) are in use.

② Plastic packaging prevents moisture loss and helps preserve weight compared to paper bags, but it also accelerates rancidity.

③ Metal-coated packaging is effective for pest control, and its effectiveness is enhanced when combined with injection of carbon dioxide (CO_2) or nitrogen (N_2).

B: Packaging units

Previously, the common packaging units were 80kg or 40kg, but now packages of 10kg or less are prevalent. Smaller packaging units are also advantageous for maintaining quality and facilitating distribution.

3. Distribution of Rice

1) Distribution Channels of Rice

① Rice production in Korea was about 5 million tons annually in 2000, but has declined to about 3.5 million tons by 2020.

② Distribution channels include four types: (i) the cooperative route centered on the National Agricultural Cooperative Federation (NACF), (ii) the general route managed by private processors and distributors, (iii) the government procurement route, and (iv) the bulk direct-consumer route.

③ Initially, rice is collected through three routes: agricultural cooperatives and local associations (52%), government procurement (14%), and private distributors (34%). From there, rice is delivered to consumers through cooperative markets, direct sales outlets, wholesale markets, and retail distributors.

2) Current Status of Distribution

(1) Distribution in Producing Areas

① Rice producers sell their harvest to government procurement agencies, milling factories, grain collectors, and Rice Processing Complexes (RPCs). RPCs, equipped with modern facilities for drying, storage, and processing, play a major role in distribution through government and cooperative procurement.

② A major issue in producing-area distribution is the lack of facilities capable of safely storing large volumes of rice harvested simultaneously in autumn.

③ Because of weak economic capacity, more than 90% of farm households must sell their rice within five months after harvest, making it difficult to hold rice for better prices.

④ Problems include excessive competition among RPCs, difficulties in securing sales channels, challenges in obtaining high-quality paddy, and cost pressures that reduce profitability.

(2) Rice Processing Complexes (RPCs)

① To improve rice quality and competitiveness, RPCs were introduced in 1991 as integrated facilities for drying, storage, milling, packaging, and distribution. After their introduction, traditional straw bag distribution shifted to bulk rice, with packaging units standardized at 20kg, 10kg, and 5kg.

② Korean RPCs emphasize processing functions but have limited storage capacity due to small silos, whereas RPCs in other countries have large silos, enabling strong storage capabilities.

③ RPCs are equipped with quality-control instruments such as equalizers, laboratory huskers, test polishers, trial dryers, moisture meters, electronic balances, automatic husking detectors, and automated inspection systems.

④ In addition to RPCs, drying and storage complexes (DSCs) are also widely in operation.

(3) Characteristics of Wholesale Distribution and Grading

① Direct marketing has increased with the expansion of e-commerce.

② The role of RPCs continues to grow, and with rising demand for high-quality rice, small-volume consumption, and efficiency, new distribution systems emphasizing quality, speed, and minimized costs are being established.

③ Direct sales from producing areas are expanding, while the function of wholesale markets is declining.

④ Since 1994, the government's rice release system for price stabilization has been replaced by an auction system, reducing the role of cooperative wholesale markets and expanding the role of the Grain Marketing Headquarters.

⑤ In wholesale rice distribution, price formation according to quality is unclear, and rice prices are often determined by production area or variety rather than objective quality.

⑥ To enhance international competitiveness, Korea needs to adopt more detailed rice quality grading. While the United States uses six grades, Korea's government procurement system distinguishes only two (pass/fail), and certified rice is classified into three grades (special, superior, and ordinary).

⑦ Korea's highest quality grade standard includes criteria such as appropriate milling yield, absence of foreign matter, complete removal of the bran layer, high transparency and luster, minimal chalkiness, uniform and plump grains, distinctive aroma, and absence of aged odors. These criteria are partly subjective.

(4) Distribution in Consumption Areas

① Since 1993, when grain trading was deregulated from a licensing to a notification system, small-scale rice shops have declined due to the low-margin strategies of large retailers. Recently, regional specialty rice, eco-friendly rice, and branded rice products have become common, with increasing online sales.

② In some cases, rice sold at final retail outlets has a moisture content as low as 12%. Even if the variety is superior and cultivation well managed, eating quality declines at such low moisture levels.

3) Changes in the Rice Distribution Environment

(1) Market Share of Distributors

① The share of private distributors declined from 80% in the 1960s to 70% in the 1980s and to around 30% in the 2000s.

② After the 1993 Uruguay Round negotiations and under the WTO regime, government procurement decreased due to subsidy reduction plans, while cooperative sales through NACF RPCs increased steadily.

(2) Factors Driving Change

① Consumer preference for high-quality, safe, convenient, and branded rice is increasing.

② Consumer influence in the rice market is growing, along with the market share and dominance of large retailers.

③ Diverse distribution channels have emerged, and competition among them has intensified. Because rice is easily standardized and purchased frequently in large quantities, online sales have expanded rapidly.

④ With market liberalization, new distribution systems emphasizing price competition, shorter distribution stages, and bulk distribution have emerged. It is expected that large retailers and internet-based markets in consumer regions will expand further.

4) Rice Distribution Technologies

(1) Distribution of High-Eating-Quality Rice

① Distribution should evaluate and classify eating quality to increase consumer preference.

② Factors directly affecting eating quality, such as moisture content, protein content, acid value, mechanical eating-quality scores, sensory evaluation scores, and milling date, should be labeled on packaging.

③ Distribution of high-eating-quality rice requires smaller package units of 2~5kg.

④ Eating quality declines with time after milling, and in large retail stores, the average distribution period of rice is 1~2 months, which is too long.

(2) Distribution of Head Rice (Whole Kernels)

① The external appearance of rice is an important factor in determining marketability.

② The first step toward high-quality rice distribution is to increase the head rice ratio. In Korea, the proportion of chalky and broken grains is the main criterion for grading rice in distribution.

③ Korea needs to strengthen its rice inspection standards and reflect quality in pricing. In the US, Japan, and China, rice is branded and priced by variety, production region, and grade, with detailed information displayed on packaging.

(3) Distribution of Branded Rice

① In Korea, a large amount of rice is marketed as blends, with limited differentiation by variety or eating quality, making it difficult to produce consistently high-quality rice.

② RPCs often blend different varieties from various regions during milling, which reduces eating quality and makes uniform quality difficult to achieve.

③ Establishing variety-specific branding systems is necessary to enhance consumer awareness of high-quality rice and to build a new distribution framework.

5) Systems for Verification of Rice Quality and Food Safety

(1) Good Agricultural Practices (GAP)

① GAP is a set of management standards for producers and managers during production and handling, ensuring safe and hygienic agricultural products.

② GAP minimizes environmental hazards and enables comprehensive management of pesticides, chemicals, heavy metals, and microorganisms during seeding, cultivation, harvest, packaging, storage, transport, and postharvest handling. It also ensures transparency so that consumers can verify management practices.

③ GAP originated in 2003, when FAO emphasized a “food chain approach” to provide consumers with safe food free from contaminants such as chemicals and microorganisms. This approach manages all stages of food, from production to consumption, systematically and transparently. As of 2020, more than 130 countries participate.

④ In Korea, GAP began in 2005 with 96 items and has since expanded to all agricultural products. By 2021, 114,000 farms (11.3%) were certified. The National Agricultural Products Quality Management Service (NAQS) oversees GAP, with certification audits delegated to private agencies. Certification applies to both farms and postharvest facilities, with 898 GAP management facilities and 62 private certification bodies.

⑤ In the United States, GAP is led by the USDA and certification is provided by the FDA, ensuring domestic food safety and supporting agricultural exports.

⑥ In the EU, only farms exceeding GAP-level practices will receive direct government payments, while Asian countries promote GAP to meet importers' food safety requirements.

(2) Traceability System (Production History System)

① The traceability system records and manages information from production to distribution and final consumption, enabling identification and corrective measures if safety issues arise.

② The system developed after the Codex Alimentarius Commission ratified the guidelines for safe handling of fruits and vegetables on July 1, 2003.

③ In Korea, pilot projects focusing on GAP, export products, and eco-friendly produce were conducted between 2003–2005, and full implementation with voluntary registration began in 2005.

④ In Japan, implementation began in 2004, with mandatory recording and transfer of production and distribution information for rice businesses starting in 2011.

(3) Hazard Analysis and Critical Control Points (HACCP)

① HACCP is a scientific and systematic approach to food safety management that identifies potential hazards in production, processing, and distribution, and establishes critical control points to eliminate or reduce these risks.

② HACCP is internationally recognized as the most effective and efficient food safety management system, strongly recommended by the United States, Japan, the European Union, and international organizations such as Codex, WHO, and FDA.

4. Rice Quality and Functionality

1) Concept of Rice Quality

(1) Rice Quality

A: Primary quality

① Refers to the external characteristics of rice such as appearance, color, grain size, and grain fullness.

② It has been of greater concern to producers (farmers) than to consumers.

③ In times of rice shortage, primary quality was considered most important, and quality was determined primarily by the government and producers.

B: Secondary quality

① Refers to characteristics as a food material, including taste, nutritional value, storability, processability, usability, and functionality.

② It has been of greater concern to consumers and processors than to producers.

③ In recent times, secondary quality has become more emphasized, and consumers now hold the initiative in determining rice quality and price.

2) Standards of Rice Quality

① While rice yield is a quantitative trait and relatively easy to evaluate objectively, rice quality is more complex and diverse.

② Although rice quality is also a quantitative trait, people's perceptions of taste differ, making standards and evaluation methods somewhat subjective.

(1) High-quality Rice Standards in Korea

① Rice must have excellent external appearance and favorable milling properties.

② After cooking, the cooked rice should appear light yellowish-white with luster, and the grains should remain intact.

③ It should have a savory aroma and taste, be sticky and elastic, and provide a soft texture when chewed.

(2) Physicochemical Characteristics of High-Quality Rice

① Good-quality rice should have a protein content of 7% or less, an amylose content of 20% or less, and an acid value ranging from 8 to 15 mg KOH per 100 g.

② Moisture content should be around 16%. Alkali digestibility value (ADV) should be relatively high, and gelatinization temperature should be medium to slightly low.

③ A higher Mg/K mineral ratio is favorable for better eating quality.

(3) Functional Rice

① Refers to rice that has been improved and cultivated to contain or be fortified with components that enhance physiological functions, such as body defense, regulation of biological rhythms, disease prevention and recovery, anti-aging, improved blood circulation, enhanced immunity, and diabetes prevention.

② It helps stabilize the nervous system or regulate the absorption of specific nutrients through biological rhythm regulation.

③ It has preventive effects against hypertension, diabetes, obesity, and malignant tumors.

④ It improves blood circulation, promotes hematopoiesis, and helps prevent arteriosclerosis.

⑤ It suppresses aging by inhibiting the formation of lipid peroxides, which are one of the causes of aging.

⑥ It enhances biological defense by reducing allergic reactions or strengthening human immunity.

3) Components of Rice Quality

(1) Appearance

① The external appearance of rice is an important factor that influences consumer purchasing behavior, though it does not fully represent intrinsic quality.

② Important appearance traits before cooking include grain shape, degree of chalkiness, transparency, color, and head rice ratio.

③ Delicious rice is typically short and round, free of chalkiness, clear, glossy, and after cooking should have a light yellowish-white appearance with luster.

(2) Eating Quality

① Eating quality refers to the taste perceived when rice is eaten and is the most direct indicator of quality, though it is difficult to measure objectively.

② It is determined comprehensively and subjectively by taste, stickiness (tactile), chewing texture (tactile), and savory aroma (olfactory).

③ It varies by region and individual and is strongly influenced by genetic traits.

④ The final evaluation is made by the consumer after cooking and tasting, making eating quality the most direct and important element of rice quality.

(3) Safety

① Each country sets maximum residue limits (MRLs) for harmful substances in food, and these standards must be observed. In Korea, pesticide toxicity and dietary intake levels are established and announced by the Ministry of Food and Drug Safety (MFDS).

② The MRL refers to the legally defined maximum concentration of pesticide residue in food considered safe for daily lifelong consumption without adverse effects.

③ For agricultural products at the production stage, residue standards announced by the Ministry of Agriculture, Food and Rural Affairs (MAFRA) are applied, while for stored or marketed products, the standards announced by the MFDS are applied.

④ In rice, MRLs are defined for 105 pesticides including 2,4-D, and for heavy metals such as cadmium (Cd).

⑤ Rice is relatively safe compared to leafy or fruit vegetables because pesticides do not directly contact the endosperm. Once the husk and bran layers are removed during milling, food safety is relatively high.

(4) Nutritional Property

① Nutritional properties such as energy content, digestibility, carbohydrate, protein, lipid, vitamins, and minerals are fundamental criteria for evaluating rice as food. In the past, higher nutrient content was regarded as higher quality, but in recent years, traits such as protein content have been seen to inversely affect eating quality.

② Functionality is now considered an important quality standard, emphasizing health-promoting components rather than simple nutrient content.

(5) Cooking Quality

① Cooking conditions such as soaking time, water absorption, cooking volume, and cooking vessel (pressure, temperature) affect rice stickiness, hardness, and cohesiveness, thereby influencing eating quality.

② Other influencing factors include water absorption rate during cooking, expansion volume, gelatinization temperature, solids dissolved in cooking water, composition of the cooking extract, and iodine absorbance.

(6) Physicochemical Properties

① Physical traits such as hardness, viscosity, gelatinization temperature, alkali digestibility, amylose content, and cohesiveness greatly affect taste and texture.

② Chemical traits such as carbohydrate, protein, lipid, mineral content, aroma, and umami components also affect flavor and sensory quality.

(7) Processability

① Milling property, marketability, and storage quality are important. Rice that undergoes less compositional loss or deterioration during storage, and that maintains high milling yield, is preferred in distribution and by traders.

Table 9-5. Components of rice quality

No.	Quality category	Components of quality
①	Appearance characteristics	Grain size and shape, transparency, degree of uniformity, chalkiness, color, freshness, head rice ratio, etc.
②	Cooking quality	Water absorption rate, expansion volume, gelatinization temperature, solids in cooking water, etc.
③	Eating quality	Glossiness of cooked rice (appearance), savory aroma, stickiness, taste, texture, physicochemical properties of rice, etc.
④	Nutritional value	Digestibility and bioavailability, protein, lipids, vitamins, functional components, minerals, and other trace elements.
⑤	Physicochemical properties	Amylose content, protein content, gelatinization temperature, Mg/K ratio, etc.
⑥	Processability	Glutinous vs. non-glutinous rice, starch composition, protein content, functional components, etc.
⑦	Safety	Mycotoxins, pesticide residues, chemical additives, etc.

② Beyond direct consumption as cooked rice, suitability for processing into rice cakes, snacks, beverages, alcoholic drinks, and noodles is also an important secondary quality trait.

4) Head Rice and Defective Rice

① Head rice refers to grains that retain the inherent varietal characteristics, are fully matured and plump, exhibit luster, and show no external defects.

② Head rice is produced when the genetic characteristics of the variety are well expressed and favorable cultivation and processing conditions are satisfied.

③ In general, the higher the proportion of head rice, the higher the overall quality of rice.

④ Defective rice refers to grains that fail to exhibit the inherent shape and color of the variety. As the proportion of defective rice increases, the appearance quality declines and eating quality deteriorates.

Table 9-6. Types of defective rice grains

No.	Type	Shape	Characteristics
①	Colored rice		Caused by fungal or bacterial infection, leading to discoloration deep into the endosperm. Spots remain even after milling. Types include spotted rice, black-spotted rice, multi-spotted rice, and reddish-brown rice.
②	Cracked rice		Grains with internal fissures. It occurs during rapid, excessive, or high-temperature drying, and more frequently in grains with low length-to-width ratio. It can also occur when grain moisture falls below 20% or when reabsorbing moisture. During milling, these grains often break into fragments.
③	Green rice		Grains with a high chlorophyll content remaining in the pericarp. Occurs in late-flowering spikelets, early harvest, excess fertilization, or lodging. Often breaks into fragments during milling.
④	Milky-white rice		Grains that appear milky white and opaque, with a cloudy surface but some gloss. Often found in weak lower spikelets. Becomes broken rice after milling. Increases under poor grain
⑤	Notched (belly) rice		Grains that are constricted in the middle. It occurs under conditions such as low temperature during grain filling, shading, excess nitrogen, phosphorus and potassium deficiency, lack of soil moisture at flowering, or poor nutrition.

⑥	Opaque rice		Grains are fully opaque and dull (different from pigmented rice) with a completely white interior. Caused by arrested development, weak spikelets, early cultivation, excessive spikelet number, or lodging. It occurs more often in *indica* rice than in temperate *japonica* rice.
⑦	White-backed rice		Common in early cultivation under high temperature during grain filling, especially in weak spikelets. It lowers the glossiness of brown rice and milling yield.
⑧	White-based rice		Occurs when nutrient accumulation in the lower part of the grain (germ region) is insufficient. It appears frequently under high temperature during grain filling, especially in the upper spikelets where starch accumulation is delayed.
⑨	White-belly rice		Grains with an opaque white center in the abdominal region. Formed due to varietal traits and environmental factors, it occurs more often in large grains or upper spikelets, under early cultivation, heavy nitrogen fertilization, additional nitrogen application, and low temperature during grain filling. Rare in *indica* varieties. Quality is nearly equivalent to head rice.
⑩	White-core rice		Grains with an opaque white center. Occurs when cell layers with poor starch accumulation collapse during drying, scattering light and making the grain appear opaque. Influenced by both genetics and environment, it appears in large grains or terminal spikelets, especially under high temperature 0~15 days after heading. These grains absorb water well and support yeast growth, making them suitable for brewing.

5) Factors Affecting Rice Quality

① The main factors influencing rice quality can be broadly classified into variety, cultivation environment, cultivation practices, and postharvest management. These factors do not act independently but are closely interrelated and exert a combined influence.

② Pre-harvest factors include variety, climate, soil, and fertilization practices, while postharvest factors include drying, storage, milling, and cooking methods.

③ In terms of eating quality, the variety exerts the greatest influence as a single factor, followed by fertilization practices. Harvest timing, drying methods, and cropping season also have significant impacts.

④ The effects of individual natural factors such as temperature, sunshine duration, and soil texture may not be large when considered separately, but their integrated impact as an environment is as significant as that of the variety itself.

(1) Varieties of High-Quality Rice

A: Characteristics of good eating-quality varieties

① Rice with good eating quality shows starch cells with fine thread-like structures after cooking and has very little protein accumulation in the starch cell walls. Since this is partly a genetic trait, the production of high-quality rice requires the selection of genetically superior eating-quality varieties.

② Generally, high-eating-quality varieties are of the *japonica* type, characterized by tall plant height, weak stems prone to lodging, and higher susceptibility to diseases. They also tend to have relatively lower yields.

B: Registered varieties

① As of the end of 2021, 339 rice varieties were registered in the National Variety List of the Korea Seed & Variety Service, but only about 30 are highly preferred in the market as high-quality varieties.

② In the past, high-yielding varieties were preferred, but now variety selection depends on whether the cultivation objective is high yield or high quality.

(2) Environmental Conditions for Producing High-Quality Rice

A: Temperature and quality

① Excessively high temperatures during grain filling increase the occurrence of cracked rice, white-backed rice, and milky-white rice, whereas low temperatures increase the incidence of immature grains, notched rice, and white-belly rice, thereby lowering quality.

② In cooler regions, accumulated temperature shows a strong correlation with the proportion of first-grade rice.

③ When temperatures are high during grain filling, especially in the early stage, protein content increases, leading to a decline in eating quality.

④ With respect to harvest timing, early harvest increases the incidence of green rice, opaque rice, and broken grains, while late harvest increases the incidence of cracked rice, white-belly rice, and damaged grains.

B: Grain filling stage and quality

① For Korean rice varieties, whether table rice or black rice, the highest yield is obtained when the grain-filling temperature is around 24°C.

② The antioxidant content of black rice is also highest at 24°C.

C: Characteristics of regions producing high-quality rice

① Climatic conditions are characterized by slightly lower average temperatures, large diurnal temperature differences, long sunshine hours, and low relative humidity. While these factors are important throughout the growth period, they are especially critical during the grain-filling stage.

② Soil conditions include low soil temperature and irrigation water temperature, higher mineral content in irrigation water, and low permeability of paddy soils.

(3) Cultivation Practices for High-Quality Rice

A: Effect of cropping season on grain quality

① Variation in rice quality according to cropping season is largely due to climatic conditions, particularly temperature and sunshine during the grain-filling period.

② Early cropping often results in high-temperature grain filling, which increases amylose content and the occurrence of cracked rice, thereby lowering grain quality.

③ To increase the proportion of head rice, cropping schedules should be adjusted so that each variety undergoes grain filling under conditions optimal for its characteristics.

B: Cropping season for high-quality rice production

① With the recent intensification of global warming, somewhat cooler temperatures are advantageous for producing high-quality rice, and therefore excessively early transplanting is not desirable for quality improvement.

② The transplanting time should be adjusted by region and variety so that grain filling occurs at the optimal temperature for best quality.

C: Fertilization practices for high-quality rice

① Nitrogen fertilization negatively affects eating quality.

② Phosphorus and magnesium fertilization generally improves eating quality.

③ When nitrogen is absorbed, protein synthesis increases and proteins accumulate in the starch cell walls.

④ Protein content in rice shows a strong negative correlation with eating quality.

⑤ Increasing nitrogen fertilization raises the incidence of white-belly and cracked rice and reduces grain transparency.

⑥ To improve eating quality, nitrogen fertilizer application should be reduced, and applied nitrogen should be managed to avoid being translocated and stored as protein in the grains after heading.

⑦ High-eating-quality table rice generally has a protein content of 7% or less in milled rice, and in recent years many varieties contain 6% or less.

⑧ A higher Mg/K ratio enhances eating quality.

⑨ Increased potassium fertilization decreases eating quality, whereas magnesium fertilization increases it.

D: Optimum harvest time

① Harvesting later than the optimum time, about 40~50 days after heading, causes declines in grain quality and eating quality proportional to the delay.

② Due to global warming, autumn temperatures are now 2–3 °C higher than in the past, which accelerates grain maturation, and thus earlier harvest is required to maintain eating quality.

③ Even when harvested at the optimum time, high grain moisture prolongs drying time and increases drying costs.

④ Drying with mechanical dryers may rapidly deteriorate grain quality.

⑤ Storing freshly harvested paddy in sacks after combine harvesting can cause significant quality loss within about 24 hours.

F: Disease and postharvest quality

① The greater the incidence of disease, the more severely eating quality deteriorates. When infection exceeds 5%, fungal spores attached to grains may remain after milling, appearing as black spots and lowering appearance quality.

② Leaf blast increases discolored grains, empty grains, and malformed grains, while panicle blast produces brown-spotted rice with odor, and sensory tests show declines in aroma, taste, and viscosity.

③ Panicle mildew reduces grain filling and lowers brown rice quality.

④ Narrow brown spot disease reduces gloss and increases brown-spotted, opaque, and chalky grains.

⑤ Bacterial leaf blight interferes with grain maturation.

⑥ Bacterial grain rot produces empty grains and brown-striped kernels.

⑦ Panicle blight causes black-spotted, multi-spotted, and brown-spotted grains, resulting in defective kernels and reduced quality.

G: Insect damage and quality

① Brown planthopper damage increases the occurrence of white-core, milky-white, and cracked rice, reducing grain quality.

② Rice water weevil damage has relatively small quality effects but increases the incidence of green and opaque rice.

③ Stink bugs and panicle nematodes cause spotted and black-spotted rice.

6) Rice Quality Evaluation and Measurement Methods

(1) Appearance Quality Evaluation

① Rice quality is evaluated by three approaches: visual inspection of external appearance, chemical analysis using instruments, and eating-quality evaluation through sensory testing.

② Although the most accurate assessment of rice quality is made by consumers eating cooked rice, external appearance is also a very important factor in market purchasing decisions.

(2) Labeling of Packaged Rice

① Packaging must indicate production year, weight, variety, origin, milling date, and the address of the producer or processor to provide consumers with information.

② Recommended but not mandatory labels include quality grades (special, superior, or ordinary). For special-grade rice, there must be no foreign matter or discolored grains, broken grains ≤ 3%, and damaged grains ≤ 0.2%.

(3) Component Analysis

① Rice quality is influenced by contents of moisture, carbohydrate, protein, lipid, ash, the properties of each component, and their interactions. Eating quality in particular is strongly affected by moisture, protein, and lipid contents.

② Near-infrared spectroscopy (NIR) is widely used to rapidly and almost real-time analyze moisture, protein, lipid, and amylose content in samples non-destructively.

7) Methods of Eating-Quality Evaluation

(1) Mechanical Evaluation of Eating Quality

① Physicochemical properties of rice can be quantified and predicted for eating-quality evaluation using regression analysis.

② Physical traits used in eating-quality evaluation include moisture content, texture measurement (texturometer), heat absorption during cooking, expansion volume, solids dissolved in cooking water, water absorption at room temperature, iodine coloration, and physical properties of cooked rice such as hardness, adhesiveness, and stickiness.

③ Major physicochemical indicators such as amylose content, protein content, ash content, amylogram characteristics, alkali digestibility, rice hardness, and solids in cooking water can estimate eating quality with 70~80% accuracy.

(2) Instrumental Measurement of Eating Quality

① Although sensory evaluation by humans is important for rice quality, it requires much time and effort and trained panelists, so instrumental methods are increasingly used.

② Mechanical taste analyzers and near-infrared spectrophotometry (NIR) evaluate eating quality non-destructively by measuring moisture, amylose, protein, and lipid contents based on regression equations, using near-infrared light of 700~2,500 nm.

③ Based on the principle that cooked rice with greater glossiness has better eating quality, some devices measure reflection and absorption of special electromagnetic waves on the rice surface.

(3) Sensory Test

① Sensory tests involve direct tasting of cooked rice. Although there can be variation among panelists, well-trained panelists can provide the most accurate evaluation of eating quality.

② Sensory testing rooms must be isolated from external noise.

③ Ventilation must be installed in aroma testing rooms to avoid interference from cooking odors.

④ Testing rooms must maintain controlled conditions: temperature 20~25°C, relative humidity 50~60%, and slightly higher pressure than the surroundings to block external odors.

⑤ Lighting should be provided by one or two white fluorescent lamps of 40~60W.

⑥ Partitions should be installed between panelists to prevent mutual influence and confusion.

(4) Sensory Test Samples

① Samples should be served at normal eating temperature, consistently across all samples, to avoid cooling and moisture loss that could cause measurement error.

② White plastic plates are commonly used, with four samples per plate including a reference sample. A measuring spoon should be used to portion the rice.

(5) Panelists

① About 20 panelists should be selected, with equal numbers of men and women and balanced age distribution, and their testing ability must be verified in advance.

② Pre-tests include threshold testing for taste sensitivity, ability to distinguish sweet (sucrose), salty (NaCl), sour (citric acid), and bitter (caffeine) solutions, consistency in repeated evaluations, and familiarity with sensory terminology.

③ Panelists must wash hands with unscented soap before tests, refrain from chewing gum or eating within 30 minutes, and follow pre-training and instructions. They must evaluate independently and avoid conversation during tests.

④ Each panelist must be presented with high-quality and low-quality samples to clearly understand and describe the corresponding sensory attributes.

(6) Cooking and Testing Conditions

① For a panel of 20, 600g of rice should be cooked under identical conditions and distributed to the panelists.

② Testing should be conducted at times with minimal influence from hunger or satiety, typically around 10 a.m. and 3 p.m. Up to four evaluations per day per person are feasible.

③ Each test uses 600 g of rice (for 20 people), washed lightly 4~5 times, and cooked under standardized conditions as in normal household preparation.

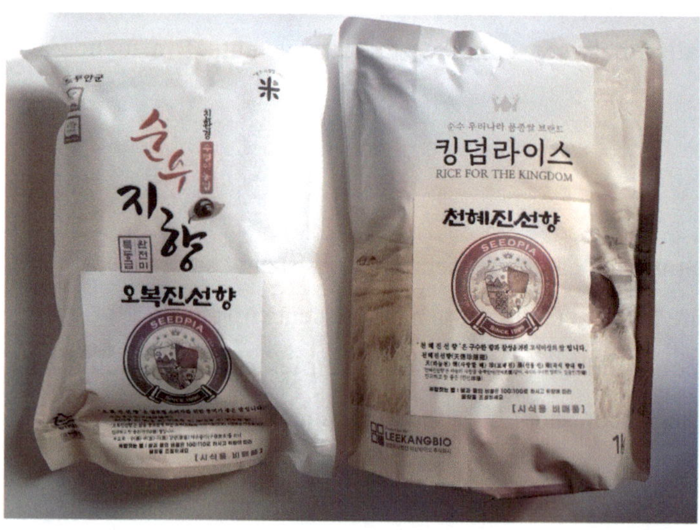

Figure 9-5.
Rice for tasting and packaging

(7) Evaluation Items and Eating-Quality Test Sheet

① The evaluation items used in sensory tests of cooked rice include appearance of the cooked rice, aroma, taste, stickiness, chewing texture, hardness, and overall rating.

② To properly evaluate whether rice quality is good or poor, a reference standard rice sample must be provided.

③ Scoring is conducted using a five-point scale, with the reference rice set as 0. Slightly better is scored +1, much better +2, slightly worse −1, and much worse −2.

8) Comprehensive Evaluation of Rice Quality

① In rice distribution, applicable quality evaluation methods combine nondestructive mechanical analysis with sensory testing. Mechanical methods are simple but less accurate, while sensory testing is practical but less objective.

② In practice, appearance quality analyzers are used to measure head-rice ratio, nondestructive analyzers to measure moisture and protein content, and taste analyzers to estimate eating quality quickly. To supplement mechanical measurement, sensory evaluation by trained panelists is also performed, allowing comprehensive assessment of rice quality.

5. Nutritional and Cooking Properties of Rice

1) Nutritional Characteristics

(1) General Composition

① Rice is nutritionally excellent, and when accompanied by various side dishes it allows balanced nutrient intake.

② The recommended energy intake ratio of carbohydrate, protein, and lipid is approximately 65:15:20 for maintaining good health.

A: Nutritional composition of rice

① Milled rice consists of about 78% starch, 7% protein, 0.4% crude fat, 0.3% crude fiber, 0.5% ash, and trace vitamins.

② Compared with brown rice, milled rice has lower contents of crude fat, crude fiber, ash, and neutral detergent fiber.

③ Ordinary milled rice contains almost no vitamin A, C, or D, but special-purpose varieties such as "Golden Rice" enriched with vitamin A have been developed, though production is limited to certain countries.

B: Carbohydrates in rice

① About 80% of rice carbohydrate is starch, which is broken down into glucose in the body, providing about 356 kcal per 100 g of milled rice.

② Since the human brain uses glucose exclusively as an energy source, rice starch is particularly important as fuel for brain activity.

③ Rice starch stimulates insulin secretion less than wheat or potato starch, helping to reduce risks of obesity and diabetes.

C: Protein in rice

① Rice protein consists mainly of 70~80% glutelin and prolamin.

② Glutelin is well digested, whereas prolamin is mostly excreted undigested.

③ Rice glutelin contains the essential amino acid lysine at about 220 mg per 100 g of milled rice, which is roughly twice as high as that of wheat flour (140 mg) and maize (110 mg).

④ Lysine helps reduce blood cholesterol levels.

⑤ Based on amino acid requirements of infants, the amino acid score of milled rice is 62, compared with 42 for wheat and 32 for maize.

⑥ Digestibility of rice protein is about 99%, compared with 96% for wheat and 95% for maize.

D: Lipids in rice (including bran)

① Brown rice contains about 2.2% lipid, while milled rice contains only 0.4%, and rice germ contains about 18% while bran contains about 17%.

② Lipids are concentrated in the bran (aleurone layer), with digestibility of about 90%.

③ Rice oil is classified as a semi-drying oil.

④ More than 70% of rice fatty acids are unsaturated, mainly oleic acid and linoleic acid, while about 20% are saturated palmitic acid, making it a high-quality oil.

⑤ Linoleic acid lowers cholesterol levels in the blood.

⑥ Lipids oxidize when exposed to air, producing volatile compounds such as hexanal and pentanal that cause rancid odors and reduce stability. However, rice lipids are essential as they provide fatty acids needed for phospholipid synthesis in cell membranes.

E: Vitamins and minerals in rice

① Rice is rich in B vitamins and contains vitamin E, folate, and niacin. The B vitamins are essential for energy release from carbohydrates but are water-soluble, so they can be lost during washing.

② Vitamin B_1, B_6, folate, vitamin B2, niacin, and pantothenic acid function as coenzymes in carbohydrate, amino acid, nucleotide, and lipid metabolism. Vitamin E (tocopherol) protects cell-membrane phospholipids from oxidative damage.

③ Minerals, abundant in bran, include phosphorus (a constituent of nucleic acids and ATP), calcium (essential for nerve and muscle function), sodium and potassium (important for nerve function and osmotic balance).

④ Iron is a component of electron transport in respiration and of hemoglobin, which carries oxygen in red blood cells.

⑤ The Mg/K ratio correlates positively with eating quality, and higher ratios improve palatability.

⑥ Dietary fiber in rice contributes to improved immunity, disease prevention, and adsorption and excretion of toxic heavy metals.

(2) Rice Bran and Germ

① Rice bran and germ, by-products of milling, contain about twice as much protein as milled rice.

② They also have much higher contents of crude fat, crude fiber, ash, and neutral detergent fiber compared with milled rice.

③ They are rich in hemicellulose, cellulose, and lignin, which are absent in milled rice.

④ Rice bran contains high levels of vitamin E and appreciable amounts of vitamins A and D.

⑤ It also contains dietary fiber, inositol, choline, pantothenic acid, tocopherols, ferulic acid, phytic acid, and other bioactive compounds.

(3) Comparison of General Composition of Rice

① Compared with major cereals, rice tends to have lower protein, fiber, and ash contents.

② Its lipid content is lower than that of barley, maize, sorghum, and oats, but higher than that of wheat and rye.

③ Compared with wheat, rice contains somewhat lower levels of protein, ash, minerals, and vitamins.

④ Rice, however, has higher essential amino acid content than wheat, particularly lysine, which is present at about twice the level, making rice especially beneficial for growing children.

⑤ Rice has high amino acid and protein quality scores, with excellent digestibility and bioavailability.

⑥ The major storage proteins differ by crop: glutelin in rice, hordein in barley, gluten in wheat, zein in maize, and globulin in soybean.

Table 9-7. Distribution of general components by rice grain part (based on milled rice moisture 14%)

(unit: %)

No.	Grain part	Carbohydrate	Lipid	Protein	Fiber	Ash
①	Bran	14	17.0	13.0	9.0	8.0
②	Brown rice	66	2.0	12.0	1.0	1.2
③	Germ	2	18.0	18.0	3.0	7.0
④	Husk (hull)	2	0.6	2.0	40.0	16.0
⑤	Milled rice (white rice)	77	0.4	7.0	0.4	0.6
⑥	Whole grain (paddy)	53	2.0	7.0	9.0	4.0

Table 9-8. Comparison of protein nutritional value in major crops

No.	Crop	Protein content (%)	Protein score	Biological value	Limiting amino acids
①	Maize (Corn)	10	42	54	Lysine, Tryptophan
②	Rice	7	72	74	Lysine, Threonine
③	Soybean	40	43	56	Methionine, Tryptophan
④	Wheat	12	47	42	Lysine, Threonine

(4) Milling and Nutrient Loss

When brown rice is milled into polished white rice, the bran and germ are removed, resulting in the loss of certain nutrients. Based on 90% milling yield (9-bun-do-mi standard), the losses are approximately: 70~80% of vitamins, 68% of protein, 57% of ash (minerals), 23% of carbohydrates, and 3.8% of lipids.

Table 9-9. Nutrient loss rate of rice by milling (unit: %)

No.	Degree of milling	Carbohydrates	Protein	Lipids	Ash	Vitamin B_1	Vitamin B_2
①	Brown rice	0	0	0	0	0	0
②	50% milled rice	11	38	1.9	26	33	30
③	70% milled rice	16	55	2.2	42	44	50
④	90% milled rice	23	68	3.8	57	77	70
⑤	120% milled rice	25	77	4.5	66	88	85

2) Cooking Characteristics

(1) Principle of Rice Cooking

A: Gelatinization

① Gelatinization is the physicochemical process in which starch granules in rice absorb water and swell upon heating, thereby increasing viscosity and producing cooked rice.

② It refers to the transformation of starch into a gelatinized state, in which the crystalline double-helical structure is disrupted and the starch becomes paste-like.

③ The conditions for gelatinization require more than 30% moisture in rice and heating at 100°C for at least 20 minutes.

④ Due to volumetric expansion during gelatinization, a single starch granule can theoretically increase up to ~60 times its original volume in sufficient water. However, in actual rice grains, the degree of expansion is only about 2.5 times because of restrictions imposed by cell walls and protein matrices.

B: Retrogradation of starch

① Retrogradation refers to the process by which gelatinized starch gradually hardens when left at room temperature, leading to a decline in palatability.

② This phenomenon occurs because starch molecules, upon losing thermal energy and water, reassociate and reform crystalline structures, causing the starch network to aggregate.

③ Retrogradation progresses rapidly in rice or rice cakes stored at low temperatures (0~3°C) with a moisture content of 30~60%, resulting in reduced palatability and digestibility.

④ Factors affecting retrogradation include starch type and concentration, the ratio of amylose to amylopectin, protein and lipid content, moisture level, temperature, pH, and the presence of salts.

⑤ Heat treatment of retrograded rice or rice cakes can reverse the process: under the influence of excited water molecules, starch is re-gelatinized.

⑥ When rice is stored for extended periods in an electric rice warmer, amino acids react with sugars to form browning compounds (Maillard reaction), causing yellow discoloration, undesirable odors, and deterioration in taste, texture, nutritional value, and digestibility.

3) Aroma of Rice

(1) Rice and Aroma Compounds

A: Volatile components constituting the aroma of raw rice

The aroma of raw rice is determined by a combination of volatile compounds, including 29 hydrocarbons, 27 aldehydes, 12 ketones, 4 alcohols, and 4 furans.

B: Aroma compounds related to rice eating quality

① The aroma of rice results from a mixture of multiple compounds, and it is influenced more by the balance of volatile compounds present in cooked rice than in raw rice.

② Characteristic volatile compounds of rice have already been identified in ordinary cooked rice.

③ A key aroma compound contributing to the "popcorn-like" flavor is not detected in raw rice but appears only in cooked rice, suggesting that it is generated by heat during cooking.

④ Since rice volatiles are primarily located in the aleurone layer, a higher milling degree reduces the amount of volatile compounds and thus weakens the aroma intensity of cooked rice.

⑤ With extended storage, linoleic acid undergoes oxidation, leading to an increase in carbonyl compounds, which causes rancid odors and deterioration of quality.

(2) Cooking Process

A: Rice cooking

① The time required for cooking rice varies depending on the quantity of rice and the type of cooker used, but typically takes 30~50 minutes.

② Gelatinization of starch in rice begins at 60~65°C.

B: Factors affecting rice cooking

① Soaking time: Proper soaking of rice in water before heating allows starch to gelatinize uniformly, improving palatability and shortening cooking time.

② Excessive soaking, however, causes leaching of nutrients, resulting in nutrient loss, reduced hardness, and increased stickiness of rice, thereby lowering eating quality.

③ At room temperature, water absorption rises sharply up to about 35% within the first 30 minutes of soaking, after which the rate increases more gradually.

④ For milled rice, an appropriate soaking time is around 30 minutes, which prevents nutrient loss, saves cooking time, and maintains palatability.

⑤ For brown rice, an adequate soaking time is 14~15 hours, due to the bran layer hindering water penetration.

⑥ Higher water temperature shortens soaking time, reduces hardness, and increases stickiness of the cooked rice.

C: Cooking water

① Complete gelatinization can be achieved with only a certain amount of water, so the volume of cooking water does not need to increase proportionally with the quantity of rice.

② The optimal cooking water amount is about 1.2 times the rice volume or 1.3~1.6 times the rice weight.

③ If excess water is added, free water decreases hardness and increases stickiness, producing overly soft rice with lower palatability. Conversely, insufficient water prevents complete gelatinization of starch, resulting in undercooked rice with higher hardness, reduced adhesiveness, and inferior eating quality.

D: Cooking vessels

① Suitable rice cookers are cast-iron pots with low thermal conductivity or thick-bottom stainless steel pots, and pressure cookers are preferable to ordinary pots.

② Cast-iron pots are advantageous because their high heat retention allows proper steaming after heating stops, and the formation of a scorched rice layer (nurungji) imparts a unique roasted aroma evenly throughout the rice.

③ A heavy lid helps maintain high internal pressure, preventing solid substances in rice grains from leaching out. This ensures rapid and uniform gelatinization, giving the rice elasticity, stickiness, and gloss.

6. Utilization and Processing of Rice

With the increase in rice production, it has now become necessary to promote utilization and processing in order to stimulate consumption. Accordingly, breeding programs for varieties suitable for processing and specific end-use characteristics are actively underway. Such specialty rices are mainly developed through mutation breeding methods. Rice is an excellent food, being more easily digested and absorbed than wheat. However, when making bread, the absence of gluten proteins and related components results in low viscoelasticity, so rice consumption in this form remains limited.

Table 9-10. Utilization of specialty rice varieties

No.	Type of specialty rice	Utilization
①	Aromatic rice	Curry rice, fried rice, lunch-box rice, rice snacks, processed cooked rice, blending rice
②	Giant-germ rice	Germ-enriched rice, germ rice dishes, nutritional supplements, health foods (notably high in GABA)
③	High-amylose rice	Rice noodles, pre-gelatinized rice, rice biscuits, culinary rice (e.g., pilaf)
④	High-globulin rice	Dietary rice for patients with allergies
⑤	Intermediate-waxy rice (semi-glutinous)	Packaged lunch-box rice, brewing, sikhye (sweet rice beverage)
⑥	Large-grain rice	Alcoholic beverages, rice snacks, soybean paste (doenjang)

⑦	Low-amylose rice	Cooked rice, processed cooked rice, pre-gelatinized rice, rice cakes (Sul-tteok), sake brewing
⑧	Low-globulin rice	Dietary rice for patients with kidney disease
⑨	Pigmented (purple) rice	Colored rice, rice cakes, alcoholic beverages, confectionery, starch syrup, natural colorant, health foods, cosmetics
⑩	Powdery rice (starchy rice)	Rice starch, rice flour
⑪	Saccharide-rich rice	Baby food, rice beverages

Table 9-11. Types and characteristics of processed rice products

No.	Processed rice product	Characteristics and uses
①	Brown rice flakes	Flaked brown rice product processed for convenient consumption, used as a substitute or snack food.
②	Extruded rice noodles	Instant noodles made by extrusion-heating rice flour into jjolmyeon or naengmyeon-type products.
③	Fermented rice beverages	Lactic acid fermented drinks and rice wine made from rice.
④	Instant dried rice (retort rice)	Ready-to-eat rice products such as retort-packed rice, canned cold rice, and dried rice requiring only hot water. Used for leisure food and emergency rations.
⑤	Instant rice	Extrusion-molded rice product that can be consumed simply by adding water.
⑥	Instant rice porridge	Ready-to-eat porridge preserving the grain shape, used as a traditional food, for patients, or as baby food.
⑦	Instant rice with meat toppings	Instant meal made by heat-processing rice together with meat and side dishes.
⑧	Misutgaru (roasted multigrain powder)	Traditional Korean powdered grain product, reformulated for easy solubility in water.
⑨	Rice-based meat (rice meat)	Meat substitute produced from rice and soybeans by extrusion molding.
⑩	Rice-flour mixed noodles	Instant noodles (e.g., rice noodles, rice ramen) made with 30~50% rice flour blend.
⑪	Sikhye (sweet rice beverage)	Traditional beverage made from rice.

Yakgwa: traditional Korean deep-fried honey pastry prepared using rice flour.	Tteokbokki rice cakes: cylindrical rice cakes made from steamed rice flour, used in spicy stir-fried rice cake dishes.	Injeolmi: rice cake made from steamed glutinous rice, coated with roasted soybean flour.
Hangwa: traditional Korean confectionery prepared from rice flour (various types, including cookies, puffs, and candies).	Tteokguk rice cakes: sliced oval-shaped rice cakes used in rice-cake soup (tteokguk).	Sulppang: ermented steamed rice bread made with yeast or rice wine.

Figure 9-6. Photographs of processed rice products

1) Cooked Rice Products

① With the launch of aseptically packaged rice that can be distributed at room temperature for extended periods, a new market for rice-based ready-to-eat meals has emerged, leading to the development of various cooked rice products.

② In rice processing, the most important factors are cooking technologies that enhance palatability and flavor-preservation technologies, both of which are strongly influenced by the packaging system.

③ Proper processing of cooked rice products has improved their convenience and storage stability.

Table 9-12. Processed cooked rice products

No.	Type of product	Description
①	Dried rice (Hwami)	Cooked and processed rice rapidly dried with hot air, enabling storage at room temperature for up to 3 years.
②	Gimbap (Triangle gimbap)	Rice wrapped in dried seaweed (gim) and shaped into rolls (or triangular form), then sliced. Shelf life is about 2~3 days.
③	Frozen rice	Cooked rice preserved by rapid freezing.
④	Freeze-dried rice	Rice preserved by freeze-drying, ready to eat simply by adding hot water.
⑤	Retort rice*	Cooked rice sealed in heat-resistant retort pouches, sterilized under pressure, and stored at room temperature. "Hetbahn", developed by CJ in 1996, was the first such product, and it is now available under various brand names.
⑥	Aseptic packaged rice	Cooked rice sealed in airtight containers, sterilized, and packaged to allow 6 months of storage at room temperature without freezing. Hetbahn products also belong to this category.
⑦	Rice balls (Jumeokbap)	Ready-to-eat rice meals made by mixing rice with beans, meat, fish, or vegetables, with binding agents and seasonings added. Shelf life about 2 days.
⑧	Instant rice with meat toppings	Instant meal made by heat-processing rice together with meat and side dishes.
⑨	Instant cup rice (Cup Rice)	Dehydrated rice that can be reconstituted by simply adding hot water is used for leisure food and emergency rations.
⑩	Canned rice	Cooked rice preserved in cans, sterilized at above 100°C, allowing up to 5 years of storage at room temperature.

<Retort Rice (Retort-Packed Rice)>

Retort rice is a processed food in which cooked rice (or other foods) is sealed in a pouch, then heated and sterilized in a high-pressure retort cooker at elevated temperatures. The pouch blocks air and light, enabling long-term preservation. The advantages compared to canned food are that it is softer in texture, lighter in weight, more convenient to transport, and can be reheated directly in its package. It becomes ready to eat after about 2 minutes of heating. If the container is flat, the rate of heat transfer during sterilization is faster, thereby shortening processing time.

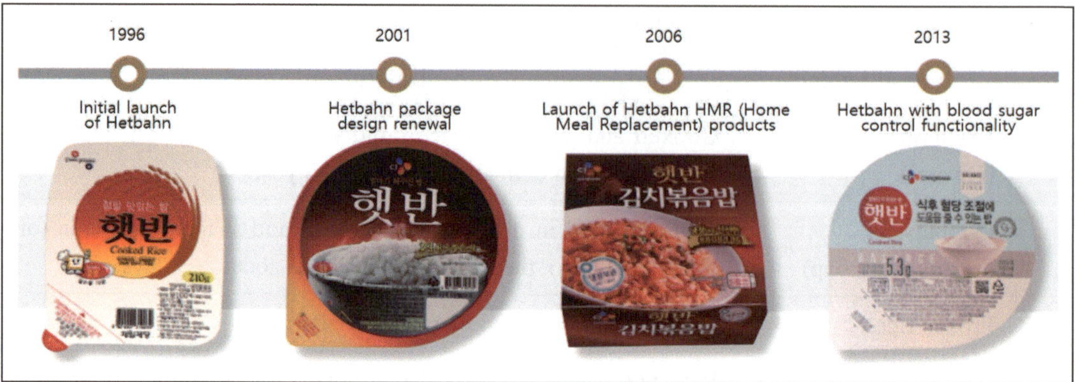

Figure 9-7. Aseptically packaged cooked rice products (e.g., Hetbahn and related items).
https://cdn.withbuyer.com/news/photo/201908/20190830_2_16533.jpg

2) Rice Cake Products (Tteok)

Rice cakes (tteok) are traditional Korean foods, believed to have been produced since the tribal state era in connection with ritual ceremonies for heaven. In Korea, tteok has long been used for festivals, ancestral rites, celebrations, staple food substitutes, and snacks.

(1) Historical records of rice cakes

① More than 240 types of rice cakes are recorded, including garaetteok, injeolmi, sirutteok, baekseolgi, pat-tteok (red bean rice cake), cinnamon tteok, and jeungpyeon (fermented rice cake).

② Steamed rice cakes (jjintteok) are the most common, and more than 100 different ingredients have been recorded as being used.

(2) Types of rice cakes

① Jjintteok (steamed rice cakes): made by steaming rice flour in a steamer (siru).

② Chintteok (pounded rice cakes): made by steaming non-glutinous or glutinous rice flour, or by steaming glutinous rice and pounding it in a mortar.

③ Jijintteok (pan-fried rice cakes): made by shaping glutinous rice flour dough and frying it in oil.

④ Salm-eun tteok (boiled rice cakes): made by shaping glutinous rice flour dough into balls and boiling them in water.

(3) Processing technology for rice cakes

① Rice cakes easily undergo retrogradation during distribution, which lowers storage stability. Traditionally, rice cakes harden within a day, making distribution difficult, but technologies have recently been developed to prevent hardening.

② Vacuum-packed white rice cakes for tteokbokki can now be distributed for over a month, and retort rice cakes that can be stored at room temperature for up to 3 months have been commercialized.

3) Rice Gruel (Juk)

Rice gruel (juk) is made by boiling rice in 5~10 times its weight of water until it acquires viscosity and fluidity. Historically, juk was considered a poor man's food, made by mixing rice with grains or vegetables and diluting it with water. Traditional forms included plain gruel (baekjuk), red bean porridge (patjuk), and pumpkin porridge (hobakjuk), while pine nut porridge (jatjuk), sesame porridge (kkaejuk), abalone porridge (jeonbokjuk), and fish porridge (eojuk) were regarded as luxury foods.

(1) Commercialization of rice gruel

① Rice gruel provides only 30~50 kcal per 100 g, which is about 30% of the caloric value of cooked rice, making it popular as a health food.

② The market for rice gruel has expanded as it is a convenient food, and it is produced not only in powdered form but also sold in retort pouches and cans.

③ Matjuk (seasoned or flavored gruel) refers to gruels combining cereals (e.g., red bean porridge) with vegetables (e.g., mugwort porridge) or shellfish (e.g., abalone porridge).

④ Instant rice gruel (instant juk) is a modern version of traditional rice gruel, with preserved grain shape and shortened cooking time.

4) Rice Noodles

(1) Rice noodles

① For rice noodle production, varieties with high amylose content and hard grain texture are desirable.

② *Japonica* rice has a low amylose content, making noodle production difficult, so it is typically blended with wheat flour.

③ *Indica* rice, due to its high amylose starch properties, can be used alone to produce noodles, and in Southeast Asia 100% rice noodles (e.g., Vietnamese rice noodles) are widely consumed.

Figure 9-8. Packaging and appearance of rice noodles

(2) Rice-Based Noodles

① Rice ramen: Instant noodles made by blending about 30% rice flour with wheat flour.

② Rice noodles: Instant noodles made by blending 40~50% rice flour.

③ Steamed noodles (jeungsuk-myeon): Instant noodles prepared by blending about 30% rice flour, steamed and dried without frying. These noodles preserve the natural flavor and texture of rice and resemble the taste of tteokguk (rice cake soup).

④ Extruded noodles: Products made from 100% rice flour, processed by extrusion and heating, and formed into noodle types such as cooking noodles, jjolmyeon (chewy noodles), and naengmyeon (cold noodles).

5) Rice Bread

When wheat flour and rice are blended in a 7:3 ratio, high-quality rice bread can be produced. Recently, with advances in baking technology, 100% rice bread has also been commercialized.

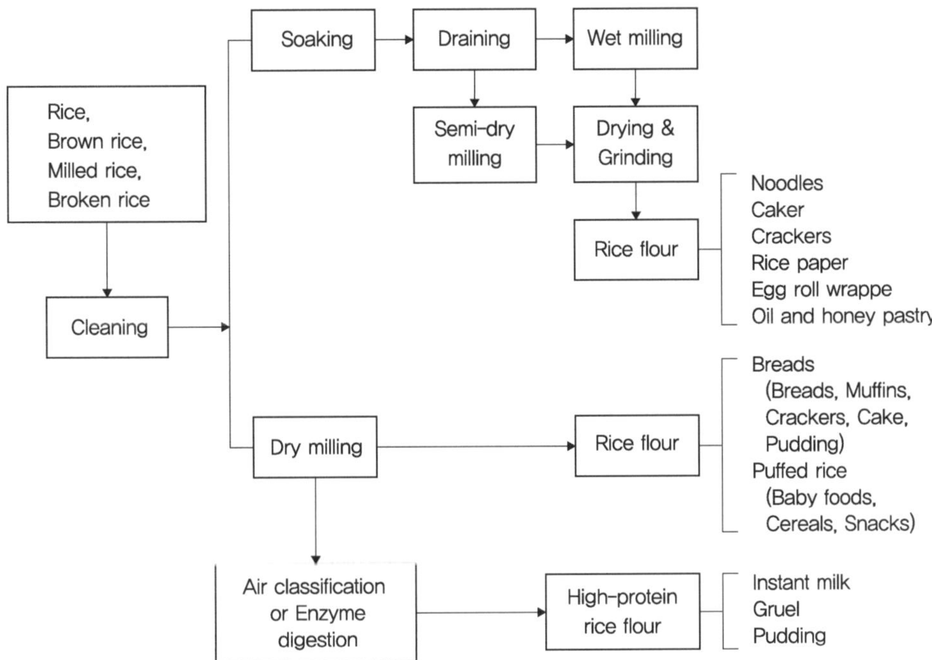

Figure 9-9. Processing flowchart of rice

6) Rice Wine and Beverages

(1) Rice Wine

① Alcoholic beverages have accompanied human history, and many are made by fermenting cooked rice with yeast to produce alcohol.

② Refined fermented rice wine yields cheongju (yakju), unrefined it is takju (makgeolli or dongdongju), and distilled it becomes soju. For brewing cheongju, large-grain rice with low protein, low lipid, low amylose content, and soft texture is preferred.

③ In Korea, more than 50 types of traditional folk rice wines, such as makgeolli, are known.

(2) Rice-Based Beverages

① Rice-based beverages include sikhye (sweet rice drink), sungnyung (toasted rice infusion), and misutgaru (roasted grain powder drink). Sikhye is made by saccharifying cooked rice with malt.

② For making refreshing beverages in the form of sweet rice drinks (gamju) or low-alcohol health beverages, large-grain glutinous rice or aromatic glutinous rice is suitable.

7) Rice Snacks

① Traditional Korean confections (hangwa) made from rice include yugwa (deep-fried rice puffs) and dasik (pressed rice cakes), as well as puffed rice products.

② Dasik is a type of hangwa that originated with tea culture during the Silla and Goryeo dynasties.

③ Yugwa is prepared by fermenting glutinous rice, grinding it into flour, mixing with soybean liquid and rice wine, rolling thin, shaping into various forms, drying, deep-frying, coating with honey or malt syrup, and finally dusting with diverse toppings.

④ Popped rice is made by placing rice into a sealed pressure vessel, heating until pressure rises, and then suddenly releasing it, causing the rice grains to expand several times their size.

⑤ Rice flakes are processed from brown rice by compression and are used as convenient staple substitutes.

8) Rice Taffy (Yeot)

① Rice taffy is made by saccharifying cooked rice with malt (yeotgireum), then boiling and concentrating it into a sticky, viscous mass.

② Depending on toppings, varieties include sesame taffy (kkaeyeot) and soybean taffy (kongyeot).

③ Depending on concentration and processing: red taffy (lightly boiled), black taffy (heavily concentrated), white taffy (aerated with bubbles), and malt syrup (jocheong) (lightly concentrated).

④ A key issue for commercialization of rice taffy is overcoming its drawback of melting at high temperatures.

9) Rice Flour

① Rice can be ground into raw rice powder (saengmibun) and used directly as starch flour, or further processed.

② Misutgaru is a soluble roasted rice flour, reformulated to dissolve even in cold water, making it a convenient portable food.

10) Rice Vinegar

① Rice vinegar contains 3~6% acetic acid, used as a sour and slightly sweet seasoning, and has been produced since the development of rice wine.

② Its consumption has increased in parallel with rising meat consumption.

③ Development of rice vinegar is directed toward health promotion, prevention of lifestyle-related diseases, cosmetic applications, and medicinal uses.

④ There is a need for microbial strains capable of producing rice vinegar with acid concentrations above 15%.

7. Utilization of By-products

1) Rice Straw

① About 4 million tons of rice straw are produced annually in Korea, nearly equal to the weight of rice production.

② Traditionally, rice straw was used to produce essential household goods such as bags (gamani), mats, baskets, ropes, and straw shoes, as well as for thatched roofs, winter feed and bedding for livestock, fuel for heating and cooking, compost, and even as a pressed material for plywood substitutes.

③ With the spread of short-stature Tongil-type rice, straw became shorter and weaker, making it unsuitable for weaving or rope-making, thus decreasing its use as a handicraft material.

④ Today, rice straw is mainly used as winter feed and bedding for livestock, as decorative material for local products, and as substrate for mushroom cultivation.

⑤ The most desirable use is returning cut straw to the paddy fields during combine harvesting to recycle organic matter.

2) Rice Husks

① Rice milling produces about 20% husk by weight.

② Botanically, husks correspond to the pericarp of the rice grain.

③ Owing to their high silica and low nitrogen content, uses are limited, and most are returned to soil as soil amendments or compost.

④ Carbonized rice husks (hwuntan) are excellent agricultural materials and were once widely used as a substrate for hydroponics.

⑤ In the past, husks were also used as stuffing for pillows, insulation material, and other household items.

3) Rice Bran

① Rice bran is produced during milling, and botanically it is a ground mixture of the seed coat (pericarp) and the aleurone layer.

② Because rice bran contains high levels of protein, lipids, minerals, vitamins, and fiber, it was historically used as a supplementary food during times of shortage.

③ It is pressed for oil, used as a preservative in pickled radish (danmuji), incorporated into compound feed, or used as a source of enzymes.

④ Rice bran contains 1~2% phytic acid ($C_6H_{18}O_{24}P_6$), which has anticancer properties and metal-chelating activity.

⑤ It is also used to prevent discoloration and spoilage of foods such as chestnuts, beans, lotus roots, burdock, pickles, and radish.

⑥ Rice bran serves as a corrosion inhibitor on metal surfaces in cans, as a stabilizer in nutritional supplements and injectables, and as an antioxidant in fragrances, cosmetics, and paints.

⑦ Rice bran contains about 20% lipids, and rice bran oil is classified as a semi-drying oil.

⑧ Its fatty acid composition is about 70% unsaturated fatty acids (oleic acid, linoleic acid) and 20% saturated fatty acids (palmitic acid).

⑨ Due to lipase activity during storage, rice bran is prone to rancidity and spoilage, and stabilization is necessary for industrial use.

⑩ Bioactive components such as tocopherols and oryzanol are extracted from rice bran and used to produce functional health foods.

Rice Crop Science

Appendix

1. Reference

Cock J.H. and Shoichi Yoshida. 1973. Photosynthesis, crop growth, and respiration of a tall and short rice varieties. Soil Sci. Plant Nutr., 19(1): 53-59.

Dobermann A. and T.H. Fairhurst. 2000. Rice: Nutrient disorders & nutrient management. International Rice Research Institute (IRRI).

Fairhurst, T.H. C. Witt, R.J. Buresh, and A. Dobermann. 2007. Rice: A Practical guide to nutrient management 2nd edition. IRRI.

Livingston B.E. and F.W. Haasis. 1933. Relations of time and maintained temperature to germination percentage for a lot of rice seed. American Journal of Botany 20: 565-619.

Maclean, J.L. D.C. Dawe, B. Hardy, and G.P. Hettel. 2002. Rice almanac. 3rd Edition. IRRI.

Nam S.Y. & K.M. Kim & J.R. Park. 2021. Food crop science. RGB.

Rebouillat J. & A. Dievart & J. L. Verdeil & J. Escoute &G. Giese & J. C. Breitler & P. Gantet & S. Espeout &E. Guiderdoni & C. Périn. 2009. Molecular Genetics of Rice Root Development. Rice 2:15–34.

Shouichi Yoshida. 1981. Fundamentals of rice crop science. IRRI.

https://www.nihhs.go.kr

https://www.kma.go.kr

https://www.mafra.go.kr/sites/mafra/index.do

https://www.rda.go.kr

http://www.fert-kfia.or.kr/

http://www.koreacpa.org/

http://www.google.co.kr

http://www.naver.com

2. Index

(N)

(O)

(P)

(R)

(S)

Rice Crop Science

First printing: February 1, 2026
First edition published: February 1, 2026

Published by: RGB Press (36cactus@naver.com)
Printed by: RGB Press (36cactus@naver.com)
ISBN 978-89-98180-47-8 / 93520

Acknowledgments

We would like to express our gratitude to Mr. Cho Heung-won, Director at Paodi, for his dedicated work in editing and printing. We also extend heartfelt thanks to the research staff of the Crop Physiology Laboratory at Sahmyook University and to our families, who patiently endured the many difficulties that arose during the preparation of this publication.

Author Biography

Sang Yong Nam (Ph.D. in Agronomy)

- B.S., M.S., and Ph.D. in Plant Production Science, College of Agriculture and Life Sciences, Seoul National University
- Professor, Department of Environmental Design & Horticulture, College of Science and Technology, Sahmyook University/Department of Environmental Horticulture, Graduate School, Sahmyook University
- Director, Natural Sciences Research Institute, Sahmyook University
- Director, Genetic Resources Management Center for Succulent Plants, National Agrobiodiversity Center, Rural Development Administration (RDA), Korea
- President, Korean Society for Cactus and Succulent Plants (KSCSP)

Kyung-Min Kim (Ph.D. in Agronomy)

- B.S., M.S., and Ph.D. in Agronomy, College of Agriculture and Life Sciences, Kyungpook National University
- Professor, School of Applied Biosciences, College of Agriculture and Life Sciences, Kyungpook National University
- Director, Coastal Agriculture Research Institute, Kyungpook National University
- Chairman, Gyeongbuk World Agriculture Forum

Jae-Ryoung Park (Ph.D. in Agronomy)

- B.S. and M.S. in Molecular Biotechnology, College of Life Resources Science, Dong-A University
- Ph.D. in Applied Biosciences, College of Agriculture and Life Sciences, Kyungpook National University
- Senior Research Scientist, National Institute of Crop Science, Rural Development Administration (RDA), Korea
- Lecturer, Department of Environmental Horticulture /Graduate School of Environmental Horticulture, Sahmyook University

Jae Hwan Lee (Ph.D. in Horticulture)

- B.S. in Environmental Design & Horticulture, College of Science and Technology, Sahmyook University
- M.S. and Ph.D. in Horticulture, Department of Environmental Horticulture, Graduate School, Sahmyook University
- Postdoctoral Researcher, Department of Environmental Horticulture, Graduate School, Sahmyook University
- Senior Researcher, Natural Science Research Institute, Sahmyook University
- Researcher, Genetic Resources Management Center for Succulent Plants, National Agrobiodiversity Center, Rural Development Administration (RDA), Korea
- Lecturer, Department of Environmental Design & Horticulture, Sahmyook University / Department of Environmental Horticulture, Graduate School, Sahmyook University